QUEUEING SYSTEMS
Problems and Solutions

QUEUEING SYSTEMS
Problems and Solutions

LEONARD KLEINROCK
Computer Science Department
School of Engineering and Applied Science
University of California, Los Angeles

RICHARD GAIL
IBM Thomas J. Watson Research Center
Yorktown Heights, New York

A Wiley-Interscience Publication

JOHN WILEY & SONS, INC.

New York / Chichester / Brisbane / Toronto / Singapore

Library of Congress Cataloging in Publication Data
Kleinrock, Leonard.
    Queueing systems : problems and solutions / Leonard Kleinrock,
  Richard Gail.
      p. cm.
    Includes index.
    ISBN 0-471-55568-1 (pbk. : alk. paper)
    1. Queueing theory. I. Gail, Richard. II. Title.
T57.92.K54   1996
519.8'2—dc20
                                    95-48333
                                        CIP

10 9 8 7

To Our Parents

Anne and Bernard Kleinrock

Helen and Harry Gail

CONTENTS

PREFACE

This book is meant to serve those who need a concise treatment of queueing theory. It is a handy guide suitable for self-study or for reference once one has mastered the topic. The book begins with a Primer in which the key concepts and results of queueing theory are presented, but not proved; this constitutes Chapter 1. Following that, the rest of the book is devoted to a statement of problems along with their detailed solutions (rather than just "answers"). These problems range from fairly straightforward exercises to rather sophisticated derivations; the latter could well be thought of as extending the material in the Primer. The problems are presented as a sequence of six chapters, each devoted to a specific topic.

These problems come from a companion text by Leonard Kleinrock, *Queueing Systems, Volume I: Theory*, Wiley-Interscience, 1975. That text provides an excellent treatment of the material contained in this book, including proofs and detailed discussion. The Primer in Chapter 1 in this book is an extended version of the same chapter in the book by Leonard Kleinrock, *Queueing Systems, Volume II: Computer Applications*, Wiley-Interscience, 1976. Moreover, the solutions are taken from the book by Leonard Kleinrock and Richard Gail, *Solutions Manual for Queueing Systems, Volume I: Theory*, Technology Transfer Institute, 1982.

The reader is encouraged to solve the problems without "peeking" at the solutions. After the reader has exhausted his/her abilities to solve the problem, the reader should then refer to the solution to check the answer or to help proceed with the solution. In all cases, it is useful to review the solutions to compare one's approach with that in this book and to study the fine points of the solution.

LEONARD KLEINROCK
RICHARD GAIL

QUEUEING SYSTEMS
Problems and Solutions

CHAPTER 1

A QUEUEING THEORY PRIMER

In this chapter we summarize the important results to which one is exposed in a first course on queueing theory. Our purpose is to lay the foundation for the theory of queueing systems. The application of this theory is to real world situations and requires sound judgment and experience in formulating models as well as in developing operational formulas (exact or approximate) that may be used for analysis and design of systems. We give a rather complete review here (by stating—*not* deriving—results) so that this material will form a self-contained body of results, which can be used in the problems and solutions that follow.

Consider any system that has a capacity C, the maximum rate at which it can perform work. Assume that R represents the average rate at which work is demanded from this system. One fundamental law of nature states that if $R < C$ then the system can "handle" the demands placed upon it, whereas if $R > C$ then the system capacity is insufficient and all the unpleasant and catastrophic effects of saturation will be experienced. However, even when $R < C$ we still experience a different set of unpleasantnesses that come about because of the *irregularity* of the demands. For example, consider the corner telephone booth, which on the average can handle the load demanded of it. Suppose now that two people approach that telephone booth almost simultaneously; it is clear that only one of the two can obtain service at a given time and the other must wait in a queue until that one is finished. Such queues arise from two sources: the first (as above) is the unscheduled arrival times of the customers; the second is the random demand (duration of service) that each customer requires of the system. The characterization of these two unpredictable quantities (the arrival times and the service times) and the evaluation of their effect on queueing phenomena form the essence of queueing theory. In the following section we introduce some of the usual notation for queueing systems and then we proceed to summarize the major results for various types of such systems.

1.1. NOTATION

Here we introduce the notation required for the statement of results in this chapter.

We consider a sequence of customers arriving to a queueing facility and let C_n denote the nth such customer to arrive. The important random variables to associate with C_n are

$$\tau_n = \text{arrival time for } C_n \tag{1.1}$$

$$t_n = \tau_n - \tau_{n-1} = \text{interarrival time between } C_n \text{ and } C_{n-1} \tag{1.2}$$

$$x_n = \text{service time for } C_n \tag{1.3}$$

It is the sequence of random variables $\{t_n\}$ and $\{x_n\}$ that really "drives" the queueing system. All these random variables are selected independently of each other, and so we define the two generic random variables

$$\tilde{t} = \text{interarrival time} \tag{1.4}$$

$$\tilde{x} = \text{service time} \tag{1.5}$$

Associated with each is a probability distribution function (PDF), that is,

$$A(t) = P[\tilde{t} \le t] \tag{1.6}$$

$$B(x) = P[\tilde{x} \le x] \tag{1.7}$$

and the related probability density function (pdf), namely,

$$a(t) = \frac{dA(t)}{dt} \tag{1.8}$$

$$b(x) = \frac{dB(x)}{dx} \tag{1.9}$$

In this last definition for the pdf we permit the use of impulse functions as discussed, for example, in [KLEI 75]. Let $E[X]$ denote the expectation (i.e., the first moment) of the random variable, X, whose pdf is $f(x)$. That is,

$$E[X] = \int_{-\infty}^{\infty} x f(x)\, dx \tag{1.10}$$

The moments associated with the random variables above are denoted by

$$E[\tilde{t}] = \bar{t} = \frac{1}{\lambda} \tag{1.11}$$

$$E[(\tilde{t})^k] = \overline{t^k} \tag{1.12}$$

$$E[\tilde{x}] = \bar{x} = \frac{1}{\mu} \tag{1.13}$$

$$E[(\tilde{x})^k] = \overline{x^k} \tag{1.14}$$

The symbol λ denotes the arrival rate of customers. The symbol μ is often reserved only for the case of exponentially distributed service times. In addition, we let σ_a^2 and σ_b^2 be the variance of $\tilde{t}$ and $\tilde{x}$, respectively.

Furthermore, we need the *Laplace transform* associated with these pdf's, namely,

$$E[e^{-s\tilde{t}}] = A^*(s) \tag{1.15}$$

$$E[e^{-s\tilde{x}}] = B^*(s) \tag{1.16}$$

We denote the relationship between a function and its transform by means of a double-barred, double-headed arrow; so, for example, $a(t) \Leftrightarrow A^*(s)$. The integral representation of this transform [say, for $a(t)$] is simply

$$A^*(s) = \int_{0^-}^{\infty} a(t)e^{-st}\, dt \tag{1.17}$$

We note that the lower limit of this integral is 0^- instead of $-\infty$, since both $\tilde{t}$ and $\tilde{x}$ are non-negative random variables.

Given a function, say, $f(t)$, it is straightforward to find its transform, $F^*(s)$, by use of Eq. (1.17). The *inverse* of a Laplace transform is simply the function that we used to create the transform; for example, from Eq. (1.17) the function $a(t)$ is the inverse of the transform $A^*(s) = E[e^{-s\tilde{t}}]$. Given $F^*(s)$, one usually finds its inverse, $f(t)$, by inspection from a table of Laplace transforms; in Table 1.1, we provide a short list of important Laplace transforms for functions $f(t)$ such that $f(t) = 0$ for $t < 0$.

TABLE 1.1   Some Laplace Transform Pairs

Function		Transform
1. $f(t) \quad t \geq 0$	$\Leftrightarrow$	$F^*(s) = \int_0^{\infty} f(t)e^{-st}\, dt$
2. Ae^{-at}		$\dfrac{A}{s+a}$
3. te^{-at}		$\dfrac{1}{(s+a)^2}$
4. $\dfrac{t^n}{n!}e^{-at}$		$\dfrac{1}{(s+a)^{n+1}}$

The Laplace transform has a number of interesting properties, some of which we list in Table 1.2.

TABLE 1.2   Some Properties of the Laplace Transform

Function		Transform
1. $f(t)$ $t \geq 0$	$\Leftrightarrow$	$F^*(s) = \int_0^\infty f(t)e^{-st}\,dt$
2. $af(t) + bg(t)$		$aF^*(s) + bG^*(s)$
3. $f(t-a)$ $(a \geq 0)$		$e^{-as}F^*(s)$
4. $e^{-at}f(t)$		$F^*(s+a)$
5. $tf(t)$		$-\dfrac{dF^*(s)}{ds}$
6. $\dfrac{f(t)}{t}$		$\int_{s_1=s}^{\infty} F^*(s_1)\,ds_1$
7. $f(t) \circledast g(t)$		$F^*(s)G^*(s)$
8. $\dfrac{df(t)}{dt}$		$sF^*(s)$
9. $\int_0^t f(u)\,du$		$\dfrac{F^*(s)}{s}$

A key use of this transform is its moment-generating property; for example, the moments $\overline{t^k}$ may be generated from $A^*(s)$ through the relationship

$$\left. \frac{d^k A^*(s)}{ds^k} \right|_{s=0} = (-1)^k \overline{t^k} \tag{1.18}$$

We often denote the kth derivative of a function $f(t)$ evaluated at $t = t_0$ by

$$\left. \frac{d^k f(t)}{dt^k} \right|_{t=t_0} = f^{(k)}(t_0) \tag{1.19}$$

Both $\tilde{t}$ and $\tilde{x}$ are the *input* random variables to the queueing system; now we must define some of the important *performance* random variables, namely, the number of customers in the system, the waiting time per customer, and the total time that a customer spends in the system; that is,

$$N(t) = \text{number of customers in system at time } t \tag{1.20}$$

$$w_n = \text{waiting time (in queue) for } C_n \tag{1.21}$$

$$s_n = \text{system time (queue plus service) for } C_n \tag{1.22}$$

We hasten to point out that sometimes w_n is referred to as the queueing time and sometimes s_n is referred to as the sojourn time or the response time. The corresponding limiting random variables (after the system has been operating infinitely long) for a stable queue are N, $\tilde{w}$, and $\tilde{s}$. As with $\tilde{t}$ and $\tilde{x}$ we may define the PDF, the pdf, the first moment, and the appropriate transform for N, $\tilde{w}$, and $\tilde{s}$ as follows:

$$P[N \leq k] \qquad P[\tilde{w} \leq y] = W(y) \qquad P[\tilde{s} \leq y] = S(y)$$

$$P[N = k] = p_k \qquad \frac{dW(y)}{dy} = w(y) \qquad \frac{dS(y)}{dy} = s(y)$$

$$E[N] = \overline{N} \qquad E[\tilde{w}] = W \qquad E[\tilde{s}] = T$$

$$E[z^N] = P(z) \qquad E[e^{-s\tilde{w}}] = W^*(s) \qquad E[e^{-s\tilde{s}}] = S^*(s)$$

Here, the *z-transform* of the discrete non-negative random variable, N, is defined as

$$P(z) = \sum_{k=0}^{\infty} p_k z^k \tag{1.23}$$

As earlier, we denote this relationship simply as $p_k \Leftrightarrow P(z)$. The *inverse* of a *z*-transform is simply the sequence of values that we used to create the transform; for example, from Eq. (1.23), the sequence p_k is the inverse of the *z*-transform $P(z)$. In Table 1.3 we provide a short list of important *z*-transforms for sequences f_n such that $f_n = 0$ for $n < 0$.

TABLE 1.3   Some z-Transform Pairs

Sequence		z-Transform
1. $f_n \quad n = 0, 1, 2, \ldots$	$\Leftrightarrow$	$F(z) = \sum_{n=0}^{\infty} f_n z^n$
2. $A\alpha^n$		$\dfrac{A}{1 - \alpha z}$
3. $n\alpha^n$		$\dfrac{\alpha z}{(1 - \alpha z)^2}$
4. $n^2 \alpha^n$		$\dfrac{\alpha z(1 + \alpha z)}{(1 - \alpha z)^3}$
5. $\dfrac{1}{m!}(n + m)(n + m - 1) \cdots (n + 1)\alpha^n$		$\dfrac{1}{(1 - \alpha z)^{m+1}}$
6. $\dfrac{1}{n!}$		e^z

The *z*-transform has a number of interesting properties, some of which we list in Table 1.4.

The *z*-transform also has the moment-generating property; for example,

$$\left. \frac{d}{dz} E[z^N] \right|_{z=1} = \overline{N} \tag{1.24}$$

TABLE 1.4   **Some Properties of the z-Transform**

Sequence	z-Transform
1. f_n $n = 0, 1, 2, \ldots$ $\Leftrightarrow$	$F(z) = \sum_{n=0}^{\infty} f_n z^n$
2. $af_n + bg_n$	$aF(z) + bG(z)$
3. f_{n+1}	$\dfrac{1}{z}[F(z) - f_0]$
4. f_{n-k} $k > 0$	$z^k F(z)$
5. nf_n	$z\dfrac{d}{dz}F(z)$
6. $n(n - 1)(n - 2)\cdots(n - m + 1)f_n$	$z^m \dfrac{d^m}{dz^m}F(z)$
7. $f_n \circledast g_n$	$F(z)G(z)$
8. $f_n - f_{n-1}$	$(1 - z)F(z)$
9. $\sum_{k=0}^{n} f_k$ $n = 0, 1, 2, \ldots$	$\dfrac{F(z)}{1 - z}$

and

$$\left. \frac{d^2}{dz^2} E[z^N] \right|_{z=1} = \overline{N^2} - \overline{N} \tag{1.25}$$

The study of queues naturally breaks into three cases: elementary queueing theory, intermediate queueing theory, and advanced queueing theory. What distinguishes these three cases are the assumptions regarding $a(t)$ and $b(x)$. In order to name the different kinds of systems we wish to discuss, a rather simple shorthand notation is used for describing queues. This involves a three-component description, A/B/m, which denotes an m-server queueing system where A and B "describe" the interarrival time distribution and service time distribution, respectively. A and B take on values from the following set of symbols, which are meant to remind the reader to which distributions they refer:

$$M = \text{exponential (i.e., Markovian)}$$

$$E_r = r\text{-stage Erlangian}$$

$$H_R = R\text{-stage Hyperexponential}$$

$$D = \text{Deterministic}$$

$$G = \text{General}$$

The well-known exponential distribution is given in Eq. (1.26) below. The r-stage Erlangian distribution, as given in Eq. (1.27) below, represents the time it takes for a customer to pass through a series of r "stages," where the time required to pass through each stage is independent and is identically distributed as an exponential random variable [see Eq. (1.26)] with parameter $r\mu$ (i.e., with mean $1/r\mu$). When used to generate an interarrival time pdf, we refer to the r stages as an "arrival box"; when used to generate a service time pdf, we refer to it as a "service box." At most one customer may be in such a box at any time. The R-stage Hyperexponential distribution, as given in Eq. (1.28) below, represents the time it takes a customer to pass through one of R "stages," where the time to pass through the ith such stage $(i = 1, 2, \ldots, R)$ is an exponentially distributed random variable with mean $1/\mu_i$, and where the ith stage is selected as the one to pass through with probability α_i $(\alpha_1 + \alpha_2 + \cdots + \alpha_R = 1)$. At most one customer is allowed to be in the system of R stages. The deterministic distribution, given in Eq. (1.29) below, describes a random variable whose value is a designated constant with probability one. Specifically, if one of these symbols were used in place of B, for example, then it would refer to the following pdf ($x \geq 0$):

$$\text{M:} \quad b(x) = \mu e^{-\mu x} \tag{1.26}$$

$$\text{E}_r: \quad b(x) = \frac{r\mu(r\mu x)^{r-1} e^{-r\mu x}}{(r-1)!} \tag{1.27}$$

$$\text{H}_R: \quad b(x) = \sum_{i=1}^{R} \alpha_i \mu_i e^{-\mu_i x} \quad \left(\sum_{i=1}^{R} \alpha_i = 1; \; \alpha_i \geq 0\right) \tag{1.28}$$

$$\text{D:} \quad b(x) = u_0(x - \bar{x}) \tag{1.29}$$

$$\text{G:} \quad b(x) \text{ is arbitrary}$$

where, in Eq. (1.29), $u_0(x - \bar{x})$ refers to a unit impulse occurring at the position $x = \bar{x}$. Any distribution is permitted when G is assumed.

Occasionally we add one or two more items to our three-component description. Specifically, when we use the notation A/B/m/K/M, the fourth component, K, refers to a system that can hold at most K customers—including the customer in service—and any arriving customers that find K in system are "lost"—that is, they disappear without entering the system; the fifth component, M, refers to a finite customer population in which M customers continually circulate through the system. Unless otherwise specified, we assume that both K and M are infinite, in which case A/B/m/K/M $\rightarrow$ A/B/m. Furthermore, unless specified otherwise, we assume that order of service is given as first-come-first-serve (FCFS).

The simplest interesting system we consider in this chapter is the M/M/1 queue in which we have exponential interarrival times, exponential service times, and a single server (see Section 1.4). The most complicated system we consider in this chapter is G/G/1 in which the exponential distributions are replaced by arbitrary distributions (see Sections 1.2 and 1.10). In this book the majority of our results apply only to the first-come-first-serve queueing discipline. Let us now proceed with our summary of results.

1.2. GENERAL RESULTS

Perhaps the most important system parameter for G/G/1 is the *utilization factor* ρ, defined as the product of the average arrival rate of customers to the system times the average service time each requires; that is,

$$\rho = \lambda \bar{x} \tag{1.30}$$

This quantity gives the fraction of time that the single server is busy and is also equal to the ratio of the rate at which work arrives to the system divided by the capacity of the system to do work, namely, R/C as discussed earlier.[†] In the multiple-server system, G/G/m, the corresponding definition is

$$\rho = \frac{\lambda \bar{x}}{m} \tag{1.31}$$

which also is equal to R/C and may be interpreted as the expected fraction of busy servers when each server has the same distribution of service time; more generally, ρ is the expected fraction of the system's capacity that is in use. In all cases a stable system (one that yields limiting distributions for the performance random variables, which are stationary and independent of the initial state) is one for which

$$0 \le \rho < 1 \tag{1.32}$$

and we note that the case $\rho = 1$ is not permitted (except in the very special situation of a D/D/m queue). As we shall see, the closer ρ approaches unity, the larger are the queues and the waiting times; it is this *average* system load that essentially determines the system performance.

The average time in system is simply related to the average service time and the average waiting time through the fundamental equation

$$T = \bar{x} + W \tag{1.33}$$

and it is the quantity W that reflects the price we must pay for sharing a given resource (the server) with other customers. Whereas ρ is the most important system parameter, it is fair to say that one of the more famous formulas from queueing theory is *Little's result*, which relates the average number in the system to the average arrival rate and the average time spent in that system, namely,

$$\bar{N} = \lambda T \tag{1.34}$$

This very important result is extremely general in its application. The corresponding

---

[†] On the average, λ customers arrive per second and each brings $\bar{x}$ sec of work for the system; thus $R = \lambda \bar{x}$ sec of work per second. The (single-server) system can perform 1 sec of work per second of elapsed time, and so $C = 1$.

result for number and time in *queue* is simply given by

$$\overline{N}_q = \lambda W \tag{1.35}$$

where $\overline{N}_q$ is merely the average queue size. Furthermore, it is true in G/G/m that these quantities are related by[†]

$$\overline{N}_q = \overline{N} - m\rho \tag{1.36}$$

We have already given one fundamental law that applies to queueing systems, namely, that $R < C$ in order for the system to be stable. A second common and general law of nature also finds its way into our analyses; it relates the rate at which accumulation within a system occurs as a function of the input and output rates to and from that system. In particular, if we consider the system state in which k customers are present and if we let

$$P_k(t) = P[N(t) = k] \tag{1.37}$$

which is merely the probability that the system state at time t is k, then, loosely stated, we have

$$\frac{dP_k(t)}{dt} = [\text{flow rate of probability into state } k \text{ at time } t]$$

$$- [\text{flow rate of probability out of state } k \text{ at time } t] \tag{1.38}$$

Equation (1.38) is simply the flow conservation law for probability. This equation, which is an example of a differential-difference equation [KLEI 75], will allow us to write down time-dependent relationships among the system probabilities in a straightforward fashion (indeed, by inspection!). These differential-difference equations are often referred to as the "equations of motion" for the system.

Now consider a stable system, for which the probability $P_k(t)$ has a limiting value (as $t \rightarrow \infty$), which we denote by p_k (this represents the fraction of time that the system will contain k customers in the steady state). In this limit, $dP_k(t)/dt = 0$ and we are left with an equation that simply and succinctly says that "flow in = flow out" for state k; the set of such equilibrium equations for all states is often referred to as the *balance equations*.

1.3. MARKOV, BIRTH–DEATH, AND POISSON PROCESSES

Before we proceed to discuss the results for elementary queueing systems it is convenient to list some of the well-known results for some simple and important random processes that form the foundation for the queueing results we shall quote. Indeed, one of the most powerful tools we have in analyzing queueing systems is that

---

[†]This follows from $T = \bar{x} + W$ and Little's result.

of Markov processes, and much of the effort in queueing theory is devoted to their solution.

A discrete-state Markov process (either discrete or continuous time) is usually referred to as a *Markov chain*. We begin with discrete-time Markov chains where X_n denotes the discrete value of the (random) process at its nth step. The defining condition for such a Markov chain is

$$P[X_n = j \mid X_{n-1} = i_{n-1}, \ldots, X_1 = i_1] = P[X_n = j \mid X_{n-1} = i_{n-1}] \qquad (1.39)$$

This is merely an expression of the fact that the present state X_{n-1} completely summarizes all of the pertinent past history so far as that history affects the future of the process. We let

$$\pi_i^{(n)} = P[X_n = i] \qquad (1.40)$$

and denote the vector of these probabilities by

$$\boldsymbol{\pi}^{(n)} = [\pi_0^{(n)}, \pi_1^{(n)}, \ldots] \qquad (1.41)$$

and moreover we denote the one-step transition probabilities for homogeneous Markov chains by

$$p_{ij} = P[X_n = j \mid X_{n-1} = i] \qquad (1.42)$$

(By homogeneous, we mean specifically that p_{ij} does not depend on the time index, n.) We collect these transition probabilities into a square matrix denoted by $\mathbf{P} = [p_{ij}]$.

For a Markov chain, we often enumerate the states by integers, beginning with 0, namely, $i = 0, 1, 2, \ldots$. It is convenient to draw state i as a circle labeled with the integer i. We then draw branches connecting two states, say, i and j, with the branch label p_{ij} as follows

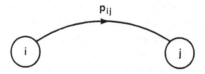

for the case of a discrete-time Markov chain. Such diagrams are called state-transition diagrams. Below, we further consider continuous-time Markov chains, in which case the branch labels are the transition rates q_{ij}; in this case the diagram is referred to as a state-transition-rate diagram. (Both state-transition diagrams and state-transition-rate diagrams are sometimes referred to simply as state diagrams.)

We have the basic results for the time-dependent probabilities of this Markov process; namely,

$$\boldsymbol{\pi}^{(n)} = \boldsymbol{\pi}^{(n-1)}\mathbf{P} \qquad (1.43)$$

$$\boldsymbol{\pi}^{(n)} = \boldsymbol{\pi}^{(0)}\mathbf{P}^n \qquad (1.44)$$

The sequence $\mathbf{P}^n$ ($n = 0, 1, 2, \ldots$) is equal to the inverse z-transform of the matrix $[\mathbf{I} - z\mathbf{P}]^{-1}$, where $\mathbf{I}$ represents the identity matrix and the superscript -1 refers to the matrix inverse.

We say that a Markov chain is *irreducible* if every state can be reached from every other state. Moreover, suppose the only possible numbers of steps at which one can return to a state after leaving it are $\gamma, 2\gamma, 3\gamma, \ldots$ (where γ is the largest such integer). If $\gamma > 1$, then that state is said to be *periodic* with period γ; if $\gamma = 1$, then that state is said to be *aperiodic*. If the probability of ever returning to a state after leaving it is unity, then that state is said to be *recurrent*; otherwise it is said to be *transient*. In the recurrent case, if the mean time to return (also known as the mean recurrence time) is finite, then that state is said to be *recurrent non-null*; otherwise, it is said to be *recurrent null*. If any state of an irreducible aperiodic Markov chain is recurrent non-null, then *all* states of that chain are recurrent non-null, and in this case (the common case of interest) the limiting probabilities

$$\pi_i = \lim_{n \to \infty} \pi_i^{(n)} \tag{1.45}$$

(also known as the steady-state probabilities) always exist, independent of the initial state probability distribution. If we let $\boldsymbol{\pi}$ be the vector of these steady-state probabilities, we may find $\boldsymbol{\pi}$ by solving the equation

$$\boldsymbol{\pi} = \boldsymbol{\pi}\mathbf{P} \tag{1.46}$$

along with the condition that

$$\sum_{i=0}^{\infty} \pi_i = 1 \tag{1.47}$$

As mentioned above, the solution of discrete-state discrete-time Markovian queueing systems is typically devoted to solving the system of equations given in Eqs. (1.46) and (1.47). We further note that this involves the solution of a system of linear equations and hence requires matrix inversion, in general. Finally, we comment that the time the process spends in any state is geometrically distributed (an inherent property of all Markov processes); this distribution is the only discrete memoryless distribution. A discrete memoryless distribution, say, $P[N = k]$, has the property

$$P[N = k + k_0 \mid N \geq k_0] = P[N = k] \tag{1.48}$$

Let us now consider the case of a discrete-state continuous-time homogeneous Markov process $X(t)$; for such a continuous-time Markov chain we have a defining property similar to Eq. (1.39); namely, for $t_n > t_{n-1}$ ($n = 1, 2, \ldots$), we require

$$P[X(t_n) \geq x \mid X(t_{n-1}) = x_{n-1}, \ldots, X(t_0) = x_0] = P[X(t_n) \geq x \mid X(t_{n-1}) = x_{n-1}] \tag{1.49}$$

The time the process spends in any state is exponentially distributed for all continuous-time Markov processes; this is the only (continuous) memoryless distribution, and it is this property that makes the analysis simple. A continuous memoryless distribution, say, $F(x)$, has the property

$$F[X \geq x + x_0 \mid X \geq x_0] = P[X \geq x] \tag{1.50}$$

We now define the transition probabilities as

$$p_{ij}(t) = P[X(s + t) = j \mid X(s) = i] \tag{1.51}$$

[By homogeneous we mean specifically that $p_{ij}(t)$ does not depend on the time index s, but only on the time difference t.] The matrix of these transition probabilities will be denoted by $\mathbf{H}(t)$, and in terms of this matrix we may express the Chapman–Kolmogorov equations as

$$\mathbf{H}(t) = \mathbf{H}(t - s)\mathbf{H}(s) \tag{1.52}$$

In a real sense $\mathbf{H}(t)$ corresponds to $\mathbf{P}^n$ and that which corresponds to $\mathbf{P}$ itself is $\mathbf{H}(\Delta t)$ (namely, the transition probabilities over an infinitesimal interval). Of more use is the matrix $\mathbf{Q} = [q_{ij}]$, referred to as the infinitesimal generator of the process or, more simply, as the transition rate matrix; it is defined by

$$\mathbf{Q} = \lim_{\Delta t \to 0} \frac{\mathbf{H}(\Delta t) - \mathbf{I}}{\Delta t} \tag{1.53}$$

(the quantity q_{ij} is often referred to as the transition rate). In terms of this matrix we may then express the time-dependent behavior of our Markov process by the equation

$$\frac{d\mathbf{H}(t)}{dt} = \mathbf{H}(t)\mathbf{Q} \tag{1.54}$$

whose solution is

$$\mathbf{H}(t) = e^{\mathbf{Q}t} \tag{1.55}$$

The steady-state behavior of this process, namely, the stable probabilities $\boldsymbol{\pi}$, are given through the basic equation

$$\boldsymbol{\pi}\mathbf{Q} = 0 \tag{1.56}$$

along with the normalizing equation (1.47). The solution of discrete-state continuous-time Markovian queueing systems is typically devoted to solving the system of linear equations given in Eqs. (1.47) and (1.56). Whereas we do not discuss them herein, we do consider the continuous-state processes in later sections. (A more complete review of Markov chains is given in tabular form in the summary of results in [KLEI 75].)

Perhaps the most fundamental random process we encounter in queueing theory is the *Poisson process*, which describes a collection of arrivals for which the interarrival times are independent and identically distributed exponential random variables. In particular, for a Poisson process with a mean interarrival time $\bar{t} = 1/\lambda$, the probability

$P_k(t)$ of k arrivals in an interval whose duration is t sec is given by

$$P_k(t) = \frac{(\lambda t)^k}{k!} e^{-\lambda t} \qquad (1.57)$$

The average number of arrivals during this interval is merely

$$\overline{N}(t) = \lambda t \qquad (1.58)$$

and the variance of this number is given by

$$\sigma^2_{N(t)} = \lambda t \qquad (1.59)$$

We note that the mean and variance for this process are identical. The z-transform for this process is simply given by

$$E[z^{N(t)}] = e^{\lambda t(z-1)} \qquad (1.60)$$

The assumption of an exponential interarrival time means, of course,

$$a(t) = \lambda e^{-\lambda t} \qquad t \geq 0 \qquad (1.61)$$

which, we repeat, is the memoryless distribution. Here, the mean and variance are, respectively, $\bar{t} = 1/\lambda$ and $\sigma^2 = 1/\lambda^2$.

If the arrival times to a queueing system are generated from a Poisson process, then the equilibrium probability, r_k, that an arriving customer *finds* k in the system upon his arrival will in fact be equal to p_k, the long-run probability of there being k customers in the system; that is, $p_k = r_k$ (see [KLEI 75], page 118). On the other hand, if we denote by d_k the equilibrium probability that a departure *leaves behind* k customers in the system, then $d_k = r_k$ if the system state $N(t)$ is permitted to change by at most one at any time (see Problem 5.6). Thus if we have unit state changes and Poisson arrivals, then we have the situation in which $p_k = r_k = d_k$.

Among the class of discrete-state continuous-time Markov processes there is the special case of birth–death processes in which the system state changes by at most one (up or down) in any infinitesimal interval. In such cases we talk about the birth rate λ_k, which is the average rate of births when the system contains k customers, and also of the death rate μ_k, which is the average rate at which deaths occur when the population is of size k. In this special case, we have $q_{k,k+1} = \lambda_k$, $q_{k,k-1} = \mu_k$, and all other rates $q_{ij} = 0$. The time-dependent behavior for such a system is essentially given in Eq. (1.55) and leads to the following differential-difference equations:

$$\frac{dP_k(t)}{dt} = -(\lambda_k + \mu_k)P_k(t) + \lambda_{k-1}P_{k-1}(t) + \mu_{k+1}P_{k+1}(t) \qquad k \geq 1$$

$$\qquad (1.62)$$

$$\frac{dP_0(t)}{dt} = -\lambda_0 P_0(t) + \mu_1 P_1(t) \qquad k = 0$$

where $P_k(t)$ is defined in Eq. (1.37). [Note that the time-dependent solution in the special case of a "pure birth" process with constant birth rate, namely, $\mu_k = 0$ and

$\lambda_k = \lambda$, is given in Eq. (1.57), i.e., the Poisson process.] The equilibrium behavior as defined in Eq. (1.56) takes on an especially simple form for this class of birth–death processes whose solution is given as follows (here we use the more usual notation p_k rather than π_k to denote the probability of having k customers in the system):

$$p_k = p_0 \prod_{i=0}^{k-1} \frac{\lambda_i}{\mu_{i+1}} \tag{1.63}$$

with the constant p_0 being evaluated through

$$p_0 = \frac{1}{1 + \sum_{k=1}^{\infty} \prod_{i=0}^{k-1} \lambda_i / \mu_{i+1}} \tag{1.64}$$

The application of this remarkably simple equilibrium solution leads us directly to the class of elementary queueing systems that we discuss in the next three sections.

1.4. THE M/M/1 QUEUE

The M/M/1 queue is the simplest interesting queueing system we present. It is the classic example and the analytical techniques required are rather elementary. Whereas these *techniques* do not carry over into more complex systems, the *behavior* of M/M/1 is in many ways similar to that observed in the more complex cases.

Since this system has a Poisson input (with an average arrival rate λ) and makes unit step changes (single service completions and single arrivals), then $p_k = r_k = d_k$. This distribution is given by

$$p_k = (1 - \rho)\rho^k \tag{1.65}$$

where $\rho = \lambda/\mu$ (recall that the average service time is $\bar{x} = 1/\mu$). Thus we immediately find that the average number in the system is given by

$$\bar{N} = \frac{\rho}{1 - \rho} \tag{1.66}$$

with variance

$$\sigma_N^2 = \frac{\rho}{(1 - \rho)^2} \tag{1.67}$$

Using Little's result and Eq. (1.33), we may immediately write down the two basic performance expressions for average delays in M/M/1:

$$W = \frac{\rho/\mu}{1 - \rho} \tag{1.68}$$

$$T = \frac{1/\mu}{1 - \rho} \tag{1.69}$$

The terms $\overline{N}$, W, and T all demonstrate the same common behavior as regards the utilization factor ρ; namely, they all behave inversely with respect to the quantity $(1 - \rho)$. This effect is dominant for M/M/1 as well as for most common queueing systems, and in the figure below we show the average time in system as a function of the utilization factor. Thus as ρ approaches unity from below, these average delays and queue sizes grow without bound! This is true of essentially every queueing system one encounters and shows the extreme price that must be paid if one is interested in running the system close to its capacity ($\rho = 1$).

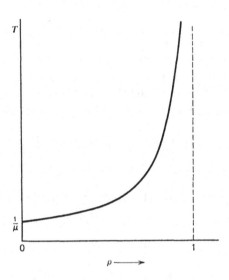

As for the distributions, we have already seen that there is a geometrically distributed number of customers in the system and we now give the waiting time and system time pdf's along with the corresponding PDF's for the case of first-come-first-serve (FCFS):

$$w(y) = (1 - \rho)u_0(y) + \lambda(1 - \rho)e^{-\mu(1-\rho)y} \qquad y \geq 0 \qquad (1.70)$$

[where we recall that $u_0(y)$ is the unit impulse (Dirac delta) function],

$$W(y) = 1 - \rho e^{-\mu(1-\rho)y} \qquad y \geq 0 \qquad (1.71)$$

$$s(y) = \mu(1 - \rho)e^{-\mu(1-\rho)y} \qquad y \geq 0 \qquad (1.72)$$

$$S(y) = 1 - e^{-\mu(1-\rho)y} \qquad y \geq 0 \qquad (1.73)$$

With the exception of the accumulation of probability at the origin for the waiting time, we note that these are all exponential in nature.

The transient behavior of most queueing systems is usually difficult to find. Such a quantity, for example, is $P_k(t)$. It turns out, for this M/M/1 system, that we can, indeed, solve for $P_k(t)$. In [KLEI 75], it can easily be shown that for

$$P_k(t) \Leftrightarrow P(z,t) = \sum_{k=0}^{\infty} P_k(t) z^k \tag{1.74}$$

and

$$P(z,t) \Leftrightarrow P^*(z,s) = \int_{0^+}^{\infty} e^{-st} P(z,t)\, dt \tag{1.75}$$

we have

$$P^*(z,s) = \frac{z^{i+1} - \mu(1-z)P_0^*(s)}{sz - (1-z)(\mu - \lambda z)} \tag{1.76}$$

where we have assumed that there are exactly i customers present in the system at $t = 0$, namely, $P_k(0) = 1$ for $k = i$ and $P_k(0) = 0$ for $k \neq i$; also $P_0(t) \Leftrightarrow P_0^*(s)$. In Problem 2.20 we find the "double" inverse of Eq. (1.76) to obtain

$$P_k(t) = e^{-(\lambda + \mu)t} \left[\rho^{(k-i)/2} I_{k-i}(at) + \rho^{(k-i-1)/2} I_{k+i+1}(at) \right.$$

$$\left. + (1-\rho)\rho^k \sum_{j=k+i+2}^{\infty} \rho^{-j/2} I_j(at) \right] \tag{1.77}$$

where $\rho = \lambda/\mu$, $a = 2\mu\rho^{1/2}$, and

$$I_k(x) \triangleq \sum_{m=0}^{\infty} \frac{(x/2)^{k+2m}}{(k+m)!\, m!} \qquad k \geq -1 \tag{1.78}$$

is the modified Bessel function of the first kind of order k.

The idle period I (the interval of time from the departure of a customer who leaves the system empty until the next arrival) and the interdeparture time D (the time between successive departures) are also both exponentially distributed with the parameter λ:

$$P[I \leq y] = P[D \leq y] = 1 - e^{-\lambda y} \qquad y \geq 0 \tag{1.79}$$

The busy period (the interval of time between successive idle periods) has a pdf denoted by $g(y)$ and is given in terms of the modified Bessel function of the first kind as

$$g(y) = \frac{1}{y\sqrt{\rho}} e^{-(\lambda+\mu)y} I_1\left(2y\sqrt{\lambda\mu}\right) \tag{1.80}$$

The probability f_n that n customers are served during a busy period is given by

$$f_n = \frac{1}{n}\binom{2n-2}{n-1} \rho^{n-1}(1+\rho)^{1-2n} \tag{1.81}$$

where the binomial coefficient is defined as

$$\binom{n}{k} = \frac{n!}{k!\,(n-k)!}$$

Two simple extensions for the M/M/1 system are easily described. First, there is the case of *bulk arrivals* where with probability g_k a group of k customers arrives at each arrival instant from the Poisson process (rather than only a single arrival); we then define the generating function for this distribution as usual by $G(z) = \sum_{k=0}^{\infty} g_k z^k$ with which we may then give the generating function for the number of customers in this bulk arrival M/M/1 system; namely,

$$P(z) = \frac{\mu(1-\rho)(1-z)}{\mu(1-z) - \lambda z[1 - G(z)]} \tag{1.82}$$

where $\rho = \lambda\bar{g}/\mu$ and $\bar{g} = G^{(1)}(1)$ is the mean size of a bulk. Note that when we have single arrivals, we have $g_1 = 1$ and $g_k = 0$ for $k \neq 1$; this gives $G(z) = z$ and then Eq. (1.82) reduces to $P(z) = (1-\rho)/(1-\rho z)$ whose inverse is given by Eq. (1.65). The second generalization is a *bulk service* system in which a free server will take up to, but no more than, r customers and serve them collectively (as if they were a single customer) with an exponentially distributed service time. The probability of finding k customers in this system is given by

$$p_k = \left(1 - \frac{1}{z_0}\right)\left(\frac{1}{z_0}\right)^k \qquad k = 0, 1, 2, \ldots \tag{1.83}$$

where z_0 is that unique root lying outside the unit disk, that is, $|z_0| > 1$, for the equation

$$r\rho z^{r+1} - (1 + r\rho)z^r + 1 = 0 \tag{1.84}$$

and where, for this case, $\rho = \lambda/r\mu$.

A final generalization involves the case of an M/M/1 system with a finite number of customers, namely, M, that behave in the following way. A customer is either in the system (waiting for service or being served) or outside the system and arriving; the interval from the time he leaves the system until he returns once again is exponentially distributed with mean $1/\lambda$. This case gives the following expression for the probability for finding k customers in the system:

$$p_k = \frac{[M!/(M-k)!](\lambda/\mu)^k}{\sum_{i=0}^{M}[M!/(M-i)!](\lambda/\mu)^i} \tag{1.85}$$

So much for the classic M/M/1 system. In the next section, we retain the Markovian assumptions but consider the case of multiple servers.

1.5. THE M/M/m QUEUEING SYSTEM

We now consider the generalization to the case of m servers. A single queue forms in front of this collection of m servers and the customer at the head of the queue will be handled by the first available server. As usual, λ is the arrival rate and $1/\mu$ is the average service time, with $\rho = \lambda/m\mu$. The equilibrium probability of finding k customers in the system is found from Eqs. (1.63) and (1.64) and is given by

$$
p_k =
\begin{cases}
p_0 \dfrac{(m\rho)^k}{k!} & k \le m \\[2ex]
p_0 \dfrac{(\rho)^k m^m}{m!} & k \ge m
\end{cases}
\tag{1.86}
$$

where

$$
p_0 = \left[\sum_{k=0}^{m-1} \frac{(m\rho)^k}{k!} + \frac{(m\rho)^m}{m!\,(1-\rho)} \right]^{-1}
\tag{1.87}
$$

A. K. Erlang, the father of queueing theory, considered this system as one model for the behavior of telephone systems early in this century [BROC 48]. Identified with his name is the Erlang-C formula, which gives the probability that an arriving customer must wait for a server; his expression is given by $\sum_{k=m}^{\infty} p_k$ from Eq. (1.86) and is often denoted by $C(m, \lambda/\mu)$.

Further results for M/M/m may be found in Section 1.9, which discusses G/M/m. Specifically W and $W(y)$ are given in Eqs. (1.137) and (1.138), respectively, where for M/M/m we have that the constant σ in those equations is simply $\sigma = \rho$.

Erlang considered a second model for telephone systems that is the same as M/M/m but which permits no customers to wait; that is, it is a loss system with at most m customers present at any one time, which, in our notation, is M/M/m/m. In this case, we again use Eqs. (1.63) and (1.64) to evaluate the probability of finding k customers in the system, which is given by

$$
p_k = \frac{(\lambda/\mu)^k / k!}{\sum_{i=0}^{m}(\lambda/\mu)^i / i!}
\tag{1.88}
$$

for the range $0 \le k \le m$. The important quantity of interest here is the probability that a customer, upon arrival to the system, will find no empty servers and will therefore be "lost"; this is referred to as the Erlang-B formula or as Erlang's Loss Formula, which is often denoted by $B(m, \lambda/\mu))$ and is given simply by p_m from Eq. (1.88).

For $m = \infty$, we have the case of an infinite server system, M/M/∞; in this system, an arriving customer always finds an available server and never waits. In this case

$$
p_k = \frac{(\lambda/\mu)^k}{k!} e^{-\lambda/\mu}
\tag{1.89}
$$

1.6. MARKOVIAN QUEUEING NETWORKS

Before leaving the comfortable world of exponential distributions, we wish to discuss another class of results that applies to *networks* of queues in which customers move from one queueing facility to another in some random fashion until they depart from the system at various points. Specifically, we consider an N-node network in which the ith node consists of a single queue served by m_i servers, each of which has an exponentially distributed service time of mean $1/\mu_i$. The ith node receives from outside the network a sequence of arrivals from an independent Poisson source at an average rate of γ_i customers per second. When a customer completes service at the ith node he will proceed next to the jth node with probability r_{ij}; thus, he becomes an "internal" arrival to the jth node. On the other hand, upon leaving the ith node a customer will depart from the entire network with probability $1 - \sum_{j=1}^{N} r_{ij}$. We define the total arrival rate to the ith node to be, on the average, λ_i customers per second, and this consists of both external and internal arrivals. The set of defining equations for λ_i is given by

$$\lambda_i = \gamma_i + \sum_{j=1}^{N} \lambda_j r_{ji} \qquad (1.90)$$

A large measure of independence exists among the nodes in these *open networks*, as may be seen from the "product-form" expression given below for the joint distribution of finding k_1 customers in the first node, k_2 customers in the second node, and so on:

$$p(k_1, k_2, \ldots, k_N) = p_1(k_1)p_2(k_2) \cdots p_N(k_N) \qquad (1.91)$$

This result is known as Jackson's theorem [JACK 57]. The factoring of this joint distribution exposes an "apparent" independence. As analysts, we are delighted to take advantage of this product form. In particular, each factor in this last expression, say, $p_i(k_i)$, is merely the solution to an *isolated* M/M/m_i queueing facility operating by itself with an input rate λ_i; the solution for $p_i(k_i)$ is given in Eq. (1.86). However, we hasten to point out that whereas the factored expression in Eq. (1.91) suggests that each node behaves as if it were receiving Poisson input, in fact, in general, the input is not Poisson.

Jackson generalized these queueing networks by allowing the total external arrival rate to the system, $\gamma(k_1 + k_2 + \cdots + k_N)$, to be a function of the total number of customers in the system. We define r_{0i} as the probability that the next external arrival goes to node i and $r_{i,N+1}$ as the probability that a customer leaving node i will depart from the network. Since the traffic rates are now functions of the total number of customers in the system, then instead of Eq. (1.90), the equivalent set is

$$e_i = r_{0i} + \sum_{j=1}^{N} e_j r_{ji} \qquad (1.92)$$

[In the case where the arrival rates are independent of the number in the system, Eqs. (1.90) and (1.92) differ by a multiplicative factor equal to the total arrival rate of customers to the system.] We assume that the solution to Eq. (1.92) exists, is unique, and is such that $e_i \geq 0$ for all i; this is equivalent to assuming that with probability one a customer's journey through the network is of finite length. The quantity e_i is, in fact, the expected number of times a customer will visit node i in passing through the network.

Another class of Markovian queueing networks (due to Gordon and Newell [GORD 67]) consists of those networks in which customers are permitted neither to leave nor to enter. In particular, we assume that K customers are placed (trapped) within a network similar to the one described above and that they move around from node to node, but no departures from the network from any node are permitted; that is, $1 - \sum_{j=1}^{N} r_{ij} = 0$ for all i. The following equations describe the behavior of the equilibrium distribution of customers in these *closed networks*:

$$p(k_1, k_2, \ldots, k_N) \sum_{i=1}^{N} \delta_{k_i - 1} \alpha_i(k_i) \mu_i$$

$$= \sum_{i=1}^{N} \sum_{j=1}^{N} \delta_{k_j - 1} \alpha_i(k_i + 1) \mu_i r_{ij} p(k_1, k_2, \ldots, k_j - 1, \ldots, k_i + 1, \ldots, k_N) \quad (1.93)$$

where the discrete unit step-function takes the form

$$\delta_k = \begin{cases} 1 & k = 0, 1, 2, \ldots \\ 0 & k < 0 \end{cases} \quad (1.94)$$

and is included in the equilibrium equations to indicate the fact that the service rate must be zero when a given node is empty; furthermore, we define

$$\alpha_i(k_i) = \begin{cases} k_i & k_i \leq m_i \\ m_i & k_i \geq m_i \end{cases} \quad (1.95)$$

which merely gives the number of customers in service in the ith node when there are k_i customers at that node. These closed networks have the following solution for the joint distribution of finding customers in various nodes:

$$p(k_1, k_2, \ldots, k_N) = \frac{1}{G(K)} \prod_{i=1}^{N} \frac{x_i^{k_i}}{\beta_i(k_i)} \quad (1.96)$$

where the set of numbers $\{x_i\}$ must satisfy the following linear equations [similar to Eq. (1.90) with $\gamma_i = 0$]:

$$\mu_i x_i = \sum_{j=1}^{N} \mu_j x_j r_{ji} \qquad i = 1, 2, \ldots, N \quad (1.97)$$

and where the normalization constant $G(K)$ guarantees conservation of probability, thus resolving the indeterminacy in the solution for $\{x_i\}$ in Eq. (1.97). The constant $G(K)$ is given by

$$G(K) = \sum_{k \in A} \prod_{i=1}^{N} \frac{x_i^{k_i}}{\beta_i(k_i)} \quad (1.98)$$

where $\mathbf{k} = (k_1, k_2, \ldots, k_N)$ and A is that set of vectors $\mathbf{k}$ for which $k_1 + k_2 + \cdots + k_N = K$ and where

$$\beta_i(k_i) = \begin{cases} k_i! & k_i \leq m_i \\ m_i! \, m_i^{k_i - m_i} & k_i \geq m_i \end{cases} \tag{1.99}$$

These open and closed networks are of considerable importance in many applications.

1.7. THE M/G/1 QUEUE

In this and the following two sections we study systems that fall in the domain of intermediate queueing theory. This classification refers to those systems in which we permit either (but not both) the interarrival time or the service time to be nonexponentially distributed; the case when both these random variables are nonexponential forms part of advanced queueing theory, which we discuss in Section 1.10. For the M/G/1 system we cannot give explicit distributions for the number in system or for the waiting time or for the time in system as we did for the M/M/1 system [specifically, see Eqs. (1.65), (1.71), and (1.73) above]. Rather, we find expressions for the transforms of their pdf's.

The M/G/1 system is characterized by a Poisson arrival process at a mean rate of λ arrivals per second and with an arbitrary or general service time distribution, $B(x)$, with a mean service time of $\bar{x}$ sec and with kth moment equal to $\overline{x^k}$. Due to the Poisson arrival process and due to the fact that the number in the system changes by at most one, we again have $p_k = r_k = d_k$.

The basic (difference) equation describing the relationship among random variables for this first-come-first-serve M/G/1 system is

$$q_{n+1} = \begin{cases} q_n - 1 + v_{n+1} & q_n > 0 \\ v_{n+1} & q_n = 0 \end{cases} \tag{1.100}$$

where q_n is the number of customers left behind by the departure of customer C_n, and v_n is the number of customers who enter during his service time (x_n). The sequence $\{q_n\}$ forms a (discrete-time) Markov chain; it is referred to as an "imbedded" Markov chain since q_n is the number in system at special points imbedded in the time axis (namely, the departure instants). The entire transient and equilibrium behavior for the system is contained in this equation, and from it we may derive most of our results for M/G/1. If we assume $\rho < 1$, then the limit $\lim_{n \to \infty} P[q_n = k]$ exists and is simply d_k, the equilibrium probability that a departure leaves behind k customers. The mean of this distribution is denoted by $\bar{q}$; however, since $p_k = d_k \, (= r_k)$ for M/G/1, then $\bar{q} = \overline{N}$.

The most well-known result for the M/G/1 system is the Pollaczek–Khinchin (P-K) mean-value formula, which gives the following compact expression for the (equilibrium) average waiting time in the queue:

$$W = \frac{\lambda \overline{x^2}/2}{1 - \rho} \tag{1.101}$$

The numerator term, denoted by $W_0 = \lambda \overline{x^2}/2$, is, in fact, equal to the expected time that a newly arriving customer must spend in the queue while that customer (if any) which he finds in service completes his remaining required service time.[†] From this formula for W one may easily calculate T using Eq. (1.33); combining that result with the result quoted in Eq. (1.34), we easily come up with the P-K mean-value formula for number in system as

$$\overline{N} = \rho + \frac{\lambda^2 \overline{x^2}/2}{1 - \rho} \qquad (1.102)$$

As mentioned above, the best we can do regarding the distributions of the various performance measures is to give the transforms associated with these random variables. The z-transform for p_k has already been defined in Eq. (1.23) as $P(z)$. We now define the z-transform for d_k as

$$Q(z) = \sum_{k=0}^{\infty} d_k z^k \qquad (1.103)$$

For M/G/1 we have

$$Q(z) = B^*(\lambda - \lambda z) \frac{(1 - \rho)(1 - z)}{B^*(\lambda - \lambda z) - z} \qquad (1.104)$$

where $B^*(\lambda - \lambda z)$ is the Laplace transform of the service time density $b(x)$ evaluated at the point $s = \lambda - \lambda z$. This last is referred to as the P-K transform equation for the number in system, and from it we easily derive Eq. (1.102). Note further that since

---

[†]This quantity is related to the concept of *residual life*, which is widely used. To elaborate, let us consider the sequence of instants (say, τ_k, $k = 0, 1, 2, \dots$) located on the real-time axis such that the set of distances between adjacent points is a set of independent, identically distributed random variables whose density we shall denote by $f(x)$ (i.e., we are dealing with a renewal process). Let m_n denote the nth moment of these interval lengths. Let us now select a point (say, t) along the time axis at random and let $t \rightarrow \infty$; the interval in which this point falls will be referred to as the "sampled" interval. The length of the sampled interval (say, X) is known as the *lifetime* of the interval, the time from the start of the sampled interval to this point is known as the *age* of the interval, and the distance (say, Y) from this selected point until the end of the sampled interval is known as the *residual life* of the interval. We are concerned with the statistics of the residual life. The pdf for residual life is given by $\hat{f}(x) = [1 - F(x)]/(m_1)$ and the Laplace transform of this density is given by $\hat{F}^*(s) = [1 - F^*(s)]/(sm_1)$; the notation here is that $F(x) = \int_0^x f(y)\,dy$ and $F^*(s)$ is the Laplace transform associated with the pdf $f(x)$. Perhaps the most significant statistic is the *mean residual life*, given by $m_2/2m_1$; that is, the expected value of the remaining length of the interval is merely the second moment over twice the first moment of the interval lengths themselves. Also, the pdf for the lifetime of the sampled interval X is $xf(x)/m_1$. One sees that W_0 is merely the mean residual life of a service time (i.e., the average remaining service time) $(\overline{x^2}/2\overline{x})$ times the probability $(\rho = \lambda\overline{x})$ that, in fact, someone is occupying the service facility.

Another quantity we wish to describe is the probability that the length of an interval (or that the value of any random variable) lies between x and $x + dx$ given that it exceeds x; dividing this probability by dx, we have a quantity referred to as the *failure rate* of the random variable, given by $f(x)/[1 - F(x)]$, where f and F refer to the pdf and the PDF of the random variable itself.

$p_k = d_k \, (= r_k)$ for M/G/1, then $P(z) = Q(z)$.[†] The Laplace transform of the waiting time pdf is merely

$$W^*(s) = \frac{s(1 - \rho)}{s - \lambda + \lambda B^*(s)} \qquad (1.105)$$

and the Laplace transform for the pdf of time in system is

$$S^*(s) = B^*(s)\frac{s(1 - \rho)}{s - \lambda + \lambda B^*(s)} \qquad (1.106)$$

These last two equations are also referred to as P-K transform equations. Due to the independence of service times, we see that Eq. (1.106) is related to Eq. (1.105) through the obvious relationship $S^*(s) = B^*(s)W^*(s)$; that is, the transform for the pdf of the sum of two independent random variables is equal to the product of the transforms of the pdf of each separately. If we divide the numerator and denominator of Eq. (1.105) by s, we then recognize the term $[1 - B^*(s)]/s\bar{x}$ (from the footnote on the previous page) as the transform of $\hat{b}(x)$, the pdf for the residual life of a service time. That is,

$$W^*(s) = \frac{1 - \rho}{1 - \rho\left[\dfrac{1 - B^*(s)}{s\bar{x}}\right]} \qquad (1.107)$$

We then invert $W^*(s)$ by inspection to find

$$w(y) = \sum_{k=0}^{\infty}(1 - \rho)\rho^k \hat{b}_{(k)}(y) \qquad y \geq 0 \qquad (1.108)$$

where $\hat{b}_0(y) = u_0(y)$ and for $k \geq 1$, $\hat{b}_{(k)}(y)$ is the k-fold convolution of the pdf $\hat{b}(y)$ with itself.

From Eq. (1.105) we easily obtain W in Eq. (1.101) by differentiation as usual; similarly, the second moment (and therefore the variance of the waiting time, denoted by σ_w^2) may be obtained to give

$$\sigma_w^2 = W^2 + \frac{\lambda \overline{x^3}}{3(1 - \rho)} \qquad (1.109)$$

Because of the Poisson arrival process, one immediately finds that the idle time I is distributed exponentially; that is,

$$P[I \leq y] = 1 - e^{-\lambda y} \qquad (1.110)$$

[†]For the case of bulk arrivals as discussed in introducing Eq. (1.82) above, the transforms $P(z)$ and $Q(z)$ are not the same. The expression for $P(z)$ is identical to that in Eq. (1.104), except that $B^*(s)$ is evaluated at the point $s = \lambda - \lambda G(z)$ rather than as above; $G(z)$ is as given just before Eq. (1.82). The expression for $Q(z)$ for the M/G/1 bulk arrival system is similar, but slightly more complicated. Expressions for both $Q(z)$ and $P(z)$ are given in Problem 5.12.

The busy-period duration has a pdf whose transform $G^*(s)$ is given through the functional equation

$$G^*(s) = B^*(s + \lambda - \lambda G^*(s)) \tag{1.111}$$

which, in general, cannot be solved. However, we may determine various moments of the busy period through the moment-generating properties of this transform, and so, for example, g_1 (the mean duration of the busy period) and σ_g^2 (the variance of this duration) are given by

$$g_1 = \frac{\overline{x}}{1 - \rho} \tag{1.112}$$

$$\sigma_g^2 = \frac{\sigma_b^2 + \rho(\overline{x})^2}{(1 - \rho)^3} \tag{1.113}$$

where σ_b^2 is the variance of the service time. Similarly, the z-transform for the number served during the busy period, which we denote by $F(z)$, is given functionally by

$$F(z) = z B^*(\lambda - \lambda F(z)) \tag{1.114}$$

with mean and variance for this number given, respectively, by

$$h_1 = \frac{1}{1 - \rho} \tag{1.115}$$

$$\sigma_h^2 = \frac{\rho(1 - \rho) + \lambda^2 \overline{x^2}}{(1 - \rho)^3} \tag{1.116}$$

An important stochastic process, which we have so far neglected, is the unfinished work, $U(t)$, in the system at time t. This is a Markov process whose value represents the time required to empty the system of all customers present at time t, assuming that no new customers enter the system after time t; that is, $U(t)$ is the system backlog expressed in time units.

For a first-come-first-serve system, the unfinished work also represents the waiting time of an arrival *if* it were to enter at time t, and so $U(t)$ is sometimes referred to as the "virtual" waiting time; in the case of a first-come-first-serve system with Poisson arrivals (M/G/1), the unfinished work has the same statistics as the true waiting time for arrivals. We wish to quote two important results regarding the distribution of $U(t)$. For this purpose, we define

$$F(w, t) = P[U(t) \leq w] \tag{1.117}$$

and we may then cite the well-known Takács integrodifferential equation, namely,

$$\frac{\partial F(w, t)}{\partial t} = \frac{\partial F(w, t)}{\partial w} - \lambda F(w, t) + \lambda \int_{x=0}^{w} B(w - x)\, d_x F(x, t) \tag{1.118}$$

which defines the transient behavior of the unfinished work distribution. Defining the double Laplace transform $F^{**}(r, s)$ for $F(w, t)$, where r carries out the transform in the w-domain and s in the t-domain, we have the following transform equation for this time-dependent behavior:

$$F^{**}(r, s) = \frac{(r/\eta)e^{-\eta w_0} - e^{-rw_0}}{\lambda B^*(r) - \lambda + r - s} \tag{1.119}$$

Here η is the unique root (for r) of the equation $s - r + \lambda - \lambda B^*(r) = 0$ in the region $\text{Re}(s) > 0$, $\text{Re}(r) > 0$, and w_0 is the initial value of the unfinished work at time 0; that is, $U(0) = w_0$.

Much more can be said about the M/G/1 system, but for purposes of this primer we have said enough. In the natural order of things we should next consider the system M/G/m, but unfortunately there are very few substantive results that can be given for this system. On the other hand, the limiting case for the M/G/∞ system is itself in some ways a trivial system since no queueing ever takes place; indeed, a very lovely result for the number of busy servers (i.e., the number of customers in the system) is given simply by

$$p_k = \frac{\rho^k}{k!}e^{-\rho} \tag{1.120}$$

We note that this result is independent of the form for $B(x)$, depending only on its first moment. Similarly, we can immediately write down that $T = \bar{x}$ and $s(y) = b(y)$. This result generalizes to the case of M/G/m/m (i.e., the system storage capacity, K, is equal to m); here we find (for $0 \leq k \leq m$)

$$p_k = \frac{\rho^k/k!}{\sum_{i=0}^{m} \rho^i/i!} \tag{1.121}$$

Once again, p_k depends only on the first moment of $B(x)$. For the (difficult) case of M/G/m with infinite storage capacity, we quote the following good approximation for the mean waiting time, $W_{M/G/m}$:

$$W_{M/G/m} \cong \frac{W_{M/M/m}}{W_{M/M/1}} \cdot W_{M/G/1} \tag{1.122}$$

1.8. THE G/M/1 QUEUE

The G/M/1 system is in fact the "dual" of the M/G/1 system. Surprisingly, G/M/1 yields to analysis more easily than M/G/1 and so we can quote distributions directly. The system, of course, corresponds to the case of an arbitrary interarrival time whose PDF is given by $A(t)$ and with pdf $a(t)$, the transform of which is denoted by $A^*(s)$; service times are distributed exponentially with mean $1/\mu$.

The basic recurrence relation that governs the behavior of G/M/1 (and also G/M/m), similar to that for M/G/1 given in Eq. (1.100), is

$$q'_{n+1} = q'_n + 1 - v'_{n+1} \qquad (1.123)$$

where q'_n is the number of customers found in the system upon the arrival of C_n and v'_{n+1} is the number of customers served between the arrival of C_n and C_{n+1}. The sequence $\{q'_n\}$ forms an imbedded Markov chain for the number in system at the arrival instants. Many of the G/M/m results follow from this equation.

All our results for G/M/1 are expressed in terms of the quantity σ that is the unique root in the range $0 \le \sigma < 1$ of the functional equation

$$\sigma = A^*(\mu - \mu\sigma) \qquad (1.124)$$

Once σ is evaluated, the following results are immediately available. The distribution for the number of customers found in the system by a new arrival is given by

$$r_k = (1 - \sigma)\sigma^k \qquad k = 0, 1, 2, \ldots \qquad (1.125)$$

The PDF for waiting time is given by

$$W(y) = 1 - \sigma e^{-\mu(1-\sigma)y} \qquad y \ge 0 \qquad (1.126)$$

and the mean waiting time is

$$W = \frac{\sigma}{\mu(1 - \sigma)} \qquad (1.127)$$

It is remarkable that the waiting times are exponentially distributed independent of the form of the interarrival time distribution (except insofar as it affects the value for σ). Note that the last three results are identical to M/M/1 with $\rho = \sigma$.

1.9. THE G/M/m QUEUE

In contrast to the M/G/m system, we find that the G/M/m system does in fact yield to analysis, the results for which we quote in this section. The G/M/m system, of course, has arbitrarily distributed interarrival times and a single queue served first-come-first-serve by m servers, each of which offers exponentially distributed service times of mean $1/\mu$. As with the system G/M/1, σ is a key parameter and in this case it is found as the unique solution in the range $0 \le \sigma < 1$ for the equation

$$\sigma = A^*(m\mu - m\mu\sigma) \qquad (1.128)$$

We have that the distribution of queue size found by a new arrival, conditioned on the fact that this arrival must queue, is given by

$$P[\text{queue size} = n \mid \text{arrival queues}] = (1 - \sigma)\sigma^n \qquad n \ge 0 \qquad (1.129)$$

We note here as with the G/M/1 system that the queue size is geometrically distributed. As earlier, we define r_k as the probability that a newly arriving customer finds k in the system ahead of him. It can be shown that

$$r_k/J = \sigma^{k-m+1} \qquad m-2 < k \tag{1.130}$$

for some J. For convenience we define $R_k = r_k/J$ $(k = 0, 1, 2, \ldots)$. We must evaluate J and the $m-1$ terms R_k for $0 \le k \le m-2$. The equation for J is given by

$$J = \frac{1}{[1/(1-\sigma)] + \sum_{k=0}^{m-2} R_k} \tag{1.131}$$

and the values for the terms R_k are given through the set of equations

$$R_{k-1} = \frac{R_k - \sum_{i=k}^{m-2} R_i p_{ik} - \sum_{i=m-1}^{\infty} \sigma^{i+1-m} p_{ik}}{p_{k-1,k}} \tag{1.132}$$

where the transition probabilities p_{ij} are nontrivial and are calculated through the following four equations, depending on the range of the subscripts i and j:

$$p_{ij} = 0 \qquad j > i+1 \tag{1.133}$$

$$p_{ij} = \int_0^\infty \binom{i+1}{j} [1 - e^{-\mu t}]^{i+1-j} e^{-\mu t j} \, dA(t) \qquad j \le i+1 \le m \tag{1.134}$$

$$p_{ij} = \int_0^\infty \frac{(m\mu t)^{i+1-j}}{(i+1-j)!} e^{-m\mu t} \, dA(t) \qquad m \le j \le i+1, \, m \le i \tag{1.135}$$

$$p_{ij} = \int_0^\infty \binom{m}{j} e^{-j\mu t} \left[\int_0^t \frac{(m\mu y)^{i-m}}{(i-m)!} (e^{-\mu y} - e^{-\mu t})^{m-j} m\mu \, dy \right] dA(t) \tag{1.136}$$

$$j < m < i+1$$

(Who said it would be easy!) Once these constants are evaluated we may then calculate the average waiting time as

$$W = \frac{J\sigma}{m\mu(1-\sigma)^2} \tag{1.137}$$

The PDF of the waiting time is given through

$$W(y) = 1 - \frac{\sigma e^{-m\mu(1-\sigma)y}}{1 + (1-\sigma)\sum_{k=0}^{m-2} R_k} \qquad y \ge 0 \tag{1.138}$$

Whereas these last two equations require the calculation of difficult constants, the waiting time pdf conditioned on the fact that the customer must queue is simply given by

$$w(y \mid \text{arrival queues}) = (1 - \sigma)m\mu e^{-m\mu(1-\sigma)y} \qquad y \geq 0 \qquad (1.139)$$

This only requires the calculation of σ. Note that even for the G/M/m system we have an exponentially distributed conditional waiting time.

1.10. THE G/G/1 QUEUE

Advanced queueing theory deals with the system G/G/1 and things beyond (e.g., G/G/m, about which we can say so very little—recall that even the system M/G/m confounded us). In this section we give some of the principal well-known results for G/G/1 and describe a method of attack that often yields the required solution or at least some simplified measures of performance. In addition we present a point of view that describes the underlying operations involved in solving the G/G/1 system.

As mentioned in the first section of this chapter, the random variables that drive any queueing system are the interarrival times t_n and the service times x_n. In the general formulation of the G/G/1 system, we find that these random variables do not appear separately in the solution but in fact always appear as a difference; thus we are led to consider a new random variable associated with the nth customer C_n, namely,

$$u_n = x_n - t_{n+1} \qquad (1.140)$$

This random variable represents the difference between the amount of work (x_n) that C_n demands of the system and the "breathing space" (t_{n+1}), or time, between the arrival of this demand and the arrival of the next demand by C_{n+1}; hopefully this difference will be negative on the average so that there will be more breathing space than load on the system. In fact if we take the average of Eq. (1.140) we find

$$E[u_n] = \bar{t}(\rho - 1) \qquad (1.141)$$

which, first of all, is independent of n (as we expected) and, second, will have a negative mean value so long as $\rho < 1$; this is no different than requiring that $R < C$ if our system is to be stable. Associated with the random variable u_n, whose generic form we now write as $\tilde{u}$, we have its PDF $C(u)$, its pdf $c(u)$, and the Laplace transform of this pdf, which we denote by $C^*(s)$. Expressing these last two in terms of the pdf's and Laplace transforms thereof for the interarrival times and service times we have

$$c(u) = \int_0^\infty b(u + t)a(t)\, dt \qquad (1.142)$$

and

$$C^*(s) = A^*(-s)B^*(s) \qquad (1.143)$$

The integral in Eq. (1.142) is, of course, the convolution integral between $a(-u)$ and $b(u)$, which we denote by $c(u) = a(-u) \circledast b(u)$. Thus once we know the interarrival time and service time pdf's we also have the pdf for our random variable $\tilde{u}$.

Of basic interest to the G/G/1 system is the behavior of the waiting time w_n for customer C_n. This random variable is related to others in the sequence through the following difference equation, in which we see the basic role played by the random variable u_n:

$$w_{n+1} = \max[0, w_n + u_n] \tag{1.144}$$

This is the key defining equation for G/G/1 [as was Eq. (1.100) for M/G/1 and Eq. (1.123) for G/M/m]. The sequence $\{w_n\}$ forms an imbedded (discrete-time continuous-state) Markov process for the unfinished work at the arrival instants. The maximum operator shown above is often rewritten in the following fashion: $(x)^+ = \max(0, x)$. In the case of a stable system ($\rho < 1$) there will exist a limiting random variable representing the equilibrium waiting time, which we denote by $\tilde{w}$. It can be seen from Eq. (1.144) that $\tilde{w}$ must have the same distribution as $(\tilde{w} + \tilde{u})^+$; the pdf that satisfies this condition will be the unique solution for the waiting time pdf. Let us denote the pdf for w_n by $w_n(y)$. The (nonlinear) functional equation that defines this pdf is given through

$$w_{n+1}(y) = \pi(w_n(y) \circledast c(y)) \tag{1.145}$$

where $\circledast$ is the convolution operator and π is a special operator (a "sweep" operator) that modifies its argument (a pdf) by replacing all of the probability associated with negative values of y (the argument of the pdf) with an impulse at $y = 0$ whose area equals this probability. The pdf $w(y)$ for our limiting random variable $\tilde{w}$ must, from Eq. (1.145), satisfy the following basic equation:

$$w(y) = \pi(w(y) \circledast c(y)) \tag{1.146}$$

whose solution will be the equilibrium density for the waiting time in G/G/1. Equation (1.146) states that this equilibrium pdf must be such that when it is convolved with $c(y)$ and when the resulting density has all of its probability on the negative half-line moved to an impulse at the origin, then we must have a resulting pdf that is the same as the $w(y)$ with which we began.

Another way to describe the random variable $\tilde{w}$ is through the equation

$$\tilde{w} = \sup_{n \geq 0} U_n \tag{1.147}$$

where $U_n = u_0 + u_1 + \cdots + u_{n-1}$ ($n \geq 1$) and $U_0 = 0$.

A random variable related to w_n that forms the "other half" for w_n is

$$y_n = -\min[0, w_n + u_n] \tag{1.148}$$

Thus we see that

$$w_{n+1} - y_n = w_n + u_n \qquad (1.149)$$

Taking expectations of this equation in the limit as $n \to \infty$, we obtain

$$\bar{y} = -\bar{u} \qquad (1.150)$$

Another defining relationship for the waiting time PDF is given by the well-known Lindley's integral equation:

$$W(y) = \begin{cases} \int_{-\infty}^{y} W(y - u) \, dC(u) & y \geq 0 \\ 0 & y < 0 \end{cases} \qquad (1.151)$$

This equation is of the Wiener–Hopf type. We now let $\Phi_+(s)$ denote the Laplace transform for the waiting time PDF $W(y)$; note that this is the transform for the PDF and not for the pdf $w(y)$, whose transform we had previously denoted by $W^*(s)$ and which is related to this new transform through the equation $W^*(s) = s\Phi_+(s)$. We wish to solve for $\Phi_+(s)$. The procedure we are about to describe is formally correct for those G/G/1 systems for which $A^*(s)$ and $B^*(s)$ may be written as rational functions of s [$A^*(s)$ and $B^*(s)$ may be approximated as closely as we like with rational functions]. In this case our task is to find a suitable representation of the following form:

$$A^*(-s)B^*(s) - 1 = \frac{\Psi_+(s)}{\Psi_-(s)} \qquad (1.152)$$

where for $\text{Re}(s) > 0$, $\Psi_+(s)$ must be an analytic function of s that contains no zeros in this half-plane; similarly, for $\text{Re}(s) < D$, $\Psi_-(s)$ must be an analytic function of s and be zero-free (where $D > 0$). In addition, we require for $|s|$ approaching infinity that the behavior of $\Psi_+(s)$ should be $\Psi_+(s) \cong s$ for $\text{Re}(s) > 0$ and that the behavior of $\Psi_-(s)$ should be $\Psi_-(s) \cong -s$ for $\text{Re}(s) < D$. Having accomplished this "spectrum factorization" we may write our solution for $\Phi_+(s)$ as

$$\Phi_+(s) = \frac{K}{\Psi_+(s)} \qquad (1.153)$$

where the constant K may be evaluated through

$$K = \lim_{s \to 0} \frac{\Psi_+(s)}{s} \qquad (1.154)$$

This constant represents the probability that an arriving customer need not queue. We note that once we have found $\Phi_+(s)$ then we have found the transform for the waiting time PDF, which is what we were seeking.

This procedure appears to be quite complicated, but, in fact, it is not. In order to show its "simplicity," let us carry it out for the case of M/M/1. Since $A^*(s) = \lambda/(s+\lambda)$ and $B^*(s) = \mu/(s + \mu)$ for M/M/1, we have

$$A^*(-s)B^*(s) - 1 = \frac{s(s - \lambda + \mu)}{(s + \mu)(\lambda - s)} \qquad (1.155)$$

In this case it was trivial to factor the numerator and the denominator. In general, this factoring is not analytically possible and we resort to numerical methods for finding the numerator roots (the "zeros" of the expression) and the denominator roots (the "poles" of the expression). This root finding is, in fact, the only hard part of the solution method; more than that, much of queueing theory often reduces to finding roots of polynomials! For this example, we plot the poles (denoted by ×'s) and the zeros (denoted by O's) in the complex s-plane as shown in the *pole-zero plot* below:

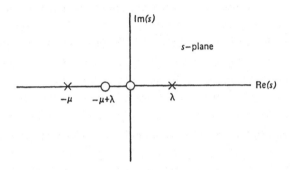

The properties of $\Psi_+(s)$ and $\Psi_-(s)$ are such that $\Psi_+(s)$ must contain *all* the poles and zeros in the region $\text{Re}(s) \le 0$ and so

$$\Psi_+(s) = \frac{s(s - \lambda + \mu)}{s + \mu} \qquad (1.156)$$

From Eq. (1.154) we have that $K = 1 - \rho$ and so, from Eq. (1.153),

$$\Phi_+(s) = \frac{(1 - \rho)(s + \mu)}{s(s - \lambda + \mu)} \qquad (1.157)$$

Inverting this expression we get the earlier result given in Eq. (1.71) for M/M/1, namely, $W(y) = 1 - \rho e^{-\mu(1-\rho)y}$ ($y \ge 0$).

Although we have described a procedure above for calculating the waiting time pdf, we have not been able to extract the properties of this solution for the general case of G/G/1 and, in fact, we have not even given an expression for the average waiting time W in the G/G/1 system. Sad to say, this quantity is, in general, unknown! Its value can be expressed, however, in terms of other system variables as follows. For example, the average waiting time is simply the negative sum of the mean residual life of the random variable $\tilde{u}$ and of $\tilde{y}$ (which is the limiting random variable for the sequence y_n); that is,

$$W = -\frac{\overline{u^2}}{2\overline{u}} - \frac{\overline{y^2}}{2\overline{y}} \qquad (1.158)$$

It can be shown that the mean residual life for $\bar{y}$ is exactly equal to the mean residual life for the random variable I, which denotes the length of an idle period in G/G/1; this last observation coupled with the easy evaluation of the first two moments of the random variable $\tilde{u}$ yields the following expression for the mean wait in G/G/1:

$$W = \frac{\sigma_a^2 + \sigma_b^2 + (\bar{\imath})^2(1 - \rho)^2}{2\bar{\imath}(1 - \rho)} - \frac{\overline{I^2}}{2\overline{I}} \tag{1.159}$$

where σ_a^2 and σ_b^2 are, respectively, the variance of the interarrival time and service time.

The waiting time transform can be expressed in terms of $I^*(s)$, the idle time transform, as

$$W^*(s) = \frac{a_0[1 - I^*(-s)]}{1 - C^*(s)} \tag{1.160}$$

where $a_0 = P[\bar{y} > 0]$. Moreover, the waiting time transform can also be expressed as

$$W'^*(s) = \frac{1 - \sigma}{1 - \sigma \hat{I}^*(s)} \tag{1.161}$$

where $\hat{I}^*(s)$ is the transform of the pdf $\hat{\imath}(y)$ corresponding to a random variable $\hat{I}$ and $\sigma = 1 - P[\tilde{w} = 0]$. Here $\hat{I}$ is the idle time in the dual queue to this G/G/1 queue, namely, a queue whose interarrival time density $\hat{a}(t) = b(t)$ and whose service time density $\hat{b}(x) = a(x)$, where $a(t)$ and $b(x)$ correspond to the original queue as in Eqs. (1.8) and (1.9). Moreover, we can show that, for G/G/1,

$$W \leq \frac{\sigma_a^2 + \sigma_b^2}{2\bar{\imath}(1 - \rho)} \tag{1.162}$$

In this book, we include no results for the G/G/m queue, whose analytic solution continues to challenge queueing theorists. An elegant approach to the exact analysis of G/G/m has been given by Kiefer and Wolfowitz [KIEF 55] involving the (often difficult) task of solving an integral equation (which reduces to Lindley's integral equation for G/G/1).

This completes our very rapid summary of the elements of queueing theory. It should be clear that a number of important behavioral properties for these queueing systems remain as yet unsolved. Nevertheless we are faced in the real world with applying the tools from queueing theory to solve immediate problems. The text [KLEI 76] addresses such problems and methods for applying the theory developed. Let us now roll up our sleeves and "attack" the problems in the remainder of this book. However, the reader is strongly urged *not* to look at the solutions until he/she has exhausted all the energy and patience at his/her disposal; only in this way will the reader benefit from this book.

REFERENCES

[BROC 48] Brockmeyer, E., H. L. Halstrøm, and A. Jensen, "The Life and Works of A. K. Erlang," *Transactions of the Danish Academy of Technology and Science*, **2**, (1948).

[CINL 75] Çinlar, E., *Introduction to Stochastic Processes*, Prentice-Hall (Englewood Cliffs, NJ), 1975.

[COHE 69] Cohen, J. W., *The Single Server Queue*, North Holland (Amsterdam), 1969.

[COX 55] Cox, D. R., "A Use of Complex Probabilities in the Theory of Stochastic Processes," *Proceedings Cambridge Philosophical Society*, **51**, 313–319 (1955).

[FELL66] Feller, W., *An Introduction to Probability Theory and Its Applications, Vol. II*, Wiley (New York), 1966.

[GORD 67] Gordon, W. J., and G. F. Newell, "Closed Queueing Systems with Exponential Servers," *Operations Research*, **15**, 254–265 (1967).

[JACK 57] Jackson, J. R., "Networks of Waiting Lines," *Operations Research*, **5**, 518–521 (1957).

[KIEF 55] Kiefer, J., and J. Wolfowitz, "On the Theory of Queues with Many Servers," *Transactions of the American Mathematical Society*, **78**, 1–18 (1955).

[KLEI 75] Kleinrock, L., *Queueing Systems, Volume I: Theory*, Wiley-Interscience (New York), 1975.

[KLEI 76] Kleinrock, L., *Queueing Systems, Volume II: Computer Applications*, Wiley-Interscience (New York), 1976.

[SAAT 61] Saaty, T. L., *Elements of Queueing Theory*, McGraw-Hill (New York), 1961.

[SYSK 60] Syski, R., *Introduction to Congestion Theory in Telephone Systems*, Oliver and Boyd (London), 1960.

[TAKA 62] Takács, L., *Introduction to the Theory of Queues*, Oxford University Press (New York), 1962.

CHAPTER 2

RANDOM PROCESSES

PROBLEM 2.1

Consider K independent sources of customers where source k is a Poisson process with rate λ_k customers per second ($k = 1, 2, \ldots, K$). Now consider the arrival stream that is formed by merging the input from each of the K sources defined above. Prove that this merged stream is also Poisson with parameter $\lambda = \lambda_1 + \lambda_2 + \cdots + \lambda_K$.

SOLUTION

Let N_k be the number of arrivals from the kth Poisson stream in an interval of duration t, and let $N = N_1 + N_2 + \cdots + N_K$. Then for $K = 2$, we have

$$P[N_1 + N_2 = n] = \sum_{j=0}^{n} P[N_1 = j] P[N_2 = n - j]$$

$$= \sum_{j=0}^{n} e^{-\lambda_1 t} \frac{(\lambda_1 t)^j}{j!} e^{-\lambda_2 t} \frac{(\lambda_2 t)^{n-j}}{(n-j)!}$$

Setting $\lambda = \lambda_1 + \lambda_2$, we obtain

$$P[N_1 + N_2 = n] = e^{-\lambda t} \frac{t^n}{n!} \sum_{j=0}^{n} \binom{n}{j} (\lambda_1)^j (\lambda_2)^{n-j}$$

$$= \frac{(\lambda t)^n}{n!} e^{-\lambda t}$$

The same argument also holds for a finite union of disjoint intervals of total duration t. This proves the result for $K = 2$ (see Theorem 1.18 in [CINL 75]). The result for general K now follows by induction. $\qquad\qquad\qquad\qquad\qquad\qquad\qquad\qquad\qquad\qquad\qquad$ □

PROBLEM 2.2

Referring back to the previous problem, consider this merged Poisson stream and now assume that we wish to break it up into several branches. Let p_i be the probability that a customer from the merged stream is assigned to the ith substream. If the overall rate is λ customers per second, and if the substream probabilities p_i are chosen for each customer independently, then show that the ith substream is a Poisson process with rate λp_i.

SOLUTION

We consider an interval of length t during which N arrivals occur from the merged stream. Let us count all the ways we can select N_i customers out of these N arrivals which join the ith substream. Since $N_i \le N$, we have

$$P[N_i = n_i] = \sum_{n = n_i}^{\infty} P[N = n] \binom{n}{n_i} p_i^{n_i} (1 - p_i)^{n - n_i}$$

$$= \sum_{n = n_i}^{\infty} e^{-\lambda t} \frac{(\lambda t)^n}{n!} \binom{n}{n_i} p_i^{n_i} (1 - p_i)^{n - n_i}$$

Thus we see that

$$P[N_i = n_i] = e^{-\lambda t} \frac{(\lambda p_i t)^{n_i}}{n_i!} \sum_{n = n_i}^{\infty} \frac{[\lambda (1 - p_i) t]^{n - n_i}}{(n - n_i)!}$$

$$= \frac{(\lambda p_i t)^{n_i}}{n_i!} e^{-\lambda p_i t}$$

The same argument also holds for a union of a finite number of disjoint intervals of total length t. This shows that the r substreams form independent Poisson processes, where the ith substream has rate λp_i (see Theorem 1.18 in [CINL 75]). $\qquad\qquad$ □

PROBLEM 2.3

Let $\{X_j\}$ be a sequence of identically distributed mutually independent Bernoulli random variables (with $P[X_j = 1] = p$, and $P[X_j = 0] = 1 - p$). Let $S_N = X_1 + \cdots + X_N$ be the sum of a random number N of the random variables X_j, where N has a Poisson distribution with mean λ. Prove that S_N has a Poisson distribution with mean λp. (In general, the distribution of the sum of a random number of independent random variables is called a compound distribution.)

SOLUTION

Condition on $N = n$. Then the conditional z-transform is

$$G(z \mid N = n) = E[z^{S_n}] = E[z^{X_1 + \cdots + X_n}] = \left(E[z^{X_1}]\right)^n$$

But

$$E[z^{X_1}] = \sum_{k=0}^{\infty} P[X_1 = k]z^k = P[X_1 = 0] \cdot 1 + P[X_1 = 1] \cdot z$$

$$= 1 - p + pz$$

Therefore

$$G(z \mid N = n) = (1 - p + pz)^n$$

Unconditioning on N, we have

$$G(z) = \sum_{n=0}^{\infty} G(z \mid N = n) \cdot P[N = n]$$

$$= \sum_{n=0}^{\infty} (1 - p + pz)^n \frac{\lambda^n e^{-\lambda}}{n!}$$

$$= e^{-\lambda} e^{\lambda(1 - p + pz)} = e^{\lambda p(z - 1)}$$

This is simply the z-transform for a Poisson distribution with parameter λp.   □

PROBLEM 2.4

Find the pdf for the smallest of K independent random variables, each of which is exponentially distributed with parameter λ.

SOLUTION

Let the K random variables be $X_1, X_2, \ldots, X_K$. The random variable of interest is $Y = \min(X_1, X_2, \ldots, X_K)$.

$$P[Y > y] = P[X_1 > y, \ldots, X_K > y]$$

$$= P[X_1 > y] \cdots P[X_K > y] \qquad (X_i \text{ are independent})$$

$$= e^{-\lambda y} \cdots e^{-\lambda y} = e^{-K\lambda y}$$

Thus Y is exponential with parameter $K\lambda$; that is,

$$P[Y \le y] = 1 - e^{-K\lambda y}$$ □

PROBLEM 2.5

Consider the homogeneous Markov chain whose state diagram is

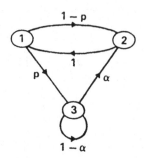

(a) Find **P**, the probability transition matrix.

(b) Under what conditions (if any) will the chain be irreducible and aperiodic?

(c) Solve for the equilibrium probability vector π.

(d) What is the mean recurrence time for state 2?

(e) For which values of α and p will we have $\pi_1 = \pi_2 = \pi_3$? (Give a physical interpretation of this case.)

SOLUTION

(a) We have

$$\mathbf{P} = \begin{bmatrix} 0 & 1-p & p \\ 1 & 0 & 0 \\ 0 & \alpha & 1-\alpha \end{bmatrix}$$

(b) Irreducible and aperiodic for all $0 < p \leq 1$ and $0 < \alpha \leq 1$ except $\alpha = p = 1$.

(c) From $\pi = \pi\mathbf{P}\,[\pi = (\pi_1, \pi_2, \pi_3)]$ we obtain only two independent equations, namely,

$$\pi_1 = \pi_2$$

$$\pi_2 = (1-p)\pi_1 + \alpha\pi_3$$

Using the conservation of probability, we also have $\pi_1 + \pi_2 + \pi_3 = 1$. Thus

$$\pi_1 = \pi_2 = \frac{\alpha}{p + 2\alpha}$$

$$\pi_3 = \frac{p}{p + 2\alpha}$$

(d) We find that

$$\mu_2 = \frac{1}{\pi_2} = \frac{p + 2\alpha}{\alpha} = 2 + \frac{p}{\alpha}$$

(e) We need

$$\alpha = p$$

Interpretation: Since each visit to state 1 is followed by exactly one visit to state 2 and vice versa for all p and α, we have $\pi_1 = \pi_2$ always. Also $p \, (= \alpha)$ of the time we go from state 1 to state 3 and the average number of steps (or mean time) spent in state 3 per visit is $1/[1 - (1 - \alpha)] = 1/\alpha$. Thus $\alpha \cdot (1/\alpha) = 1$ is the average number of visits to state 3 per visit to state 1.   □

PROBLEM 2.6

Consider the discrete-state, discrete-time Markov chain whose transition probability matrix is given by

$$\mathbf{P} = \begin{bmatrix} \frac{1}{2} & \frac{1}{2} \\ \frac{3}{4} & \frac{1}{4} \end{bmatrix}$$

(a) Find the stationary state probability vector $\boldsymbol{\pi}$.
(b) Find $[\mathbf{I} - z\mathbf{P}]^{-1}$.
(c) Find the general form for $\mathbf{P}^n$.

SOLUTION

(a) $\boldsymbol{\pi} = \boldsymbol{\pi}\mathbf{P} \, [\boldsymbol{\pi} = (\pi_1, \pi_2)]$ and $\pi_1 + \pi_2 = 1$. So

$$\pi_1 = \tfrac{3}{5}, \quad \pi_2 = \tfrac{2}{5}$$

(b) We have

$$\mathbf{I} - z\mathbf{P} = \begin{bmatrix} 1 - \frac{1}{2}z & -\frac{1}{2}z \\ -\frac{3}{4}z & 1 - \frac{1}{4}z \end{bmatrix}$$

$$\det(\mathbf{I} - z\mathbf{P}) = \left(1 - \tfrac{1}{2}z\right)\left(1 - \tfrac{1}{4}z\right) - \left(-\tfrac{1}{2}z\right)\left(-\tfrac{3}{4}z\right)$$

$$= 1 - \tfrac{3}{4}z - \tfrac{1}{4}z^2$$

$$= (1 - z)\left(1 + \tfrac{1}{4}z\right)$$

$$(\mathbf{I} - z\mathbf{P})^{-1} = \frac{1}{(1 - z)\left(1 + \frac{1}{4}z\right)} \begin{bmatrix} 1 - \frac{1}{4}z & \frac{1}{2}z \\ \frac{3}{4}z & 1 - \frac{1}{2}z \end{bmatrix}$$

(c) Using partial fraction expansions (see, e.g., [KLEI 75])

$$(I - zP)^{-1} = \frac{1}{1-z} \begin{bmatrix} \frac{3}{5} & \frac{2}{5} \\ \frac{3}{5} & \frac{2}{5} \end{bmatrix} + \frac{1}{1+\frac{1}{4}z} \begin{bmatrix} \frac{2}{5} & -\frac{2}{5} \\ -\frac{3}{5} & \frac{3}{5} \end{bmatrix}$$

Since $(I - zP)^{-1} \Leftrightarrow P^n$ we get

$$P^n = \begin{bmatrix} \frac{3}{5} & \frac{2}{5} \\ \frac{3}{5} & \frac{2}{5} \end{bmatrix} + \left(-\frac{1}{4}\right)^n \begin{bmatrix} \frac{2}{5} & -\frac{2}{5} \\ -\frac{3}{5} & \frac{3}{5} \end{bmatrix}$$ $\square$

PROBLEM 2.7

Consider a discrete-time Markov chain with transition probabilities

$$p_{ij} = e^{-\lambda} \sum_{n=0}^{j} \binom{i}{n} p^n q^{i-n} \frac{\lambda^{j-n}}{(j-n)!}$$

where $p + q = 1 \ (0 < p < 1)$.

(a) Is this chain irreducible? Periodic? Explain.

(b) We wish to find

$$\pi_i = \text{equilibrium probability of state } i$$

Write π_i in terms of p_{ij} and π_j for $j = 0, 1, 2, \ldots$.

(c) From (b) find an expression relating $P(z)$ to $P[1 + p(z - 1)]$, where

$$P(z) = \sum_{i=0}^{\infty} \pi_i z^i$$

(d) Recursively (i.e.. repeatedly) apply the result in (c) to itself and show that the nth recursion gives

$$P(z) = e^{\lambda(z-1)(1+p+p^2+\cdots+p^{n-1})} P[1 + p^n(z - 1)]$$

(e) From (d) find $P(z)$ and then recognize π_i.

SOLUTION

(a) The given Markov chain is irreducible since $p_{ij} > 0$ for all i, j (since $0 < p < 1$). It is aperiodic since $p_{ii} > 0$ for at least one (and, in fact, all) i.

(b) We have

$$\pi_i = \sum_{j=0}^{\infty} \pi_j p_{ji}$$

(c) The z-transform is

$$P(z) = \sum_{i=0}^{\infty} \pi_i z^i = \sum_{i=0}^{\infty} \sum_{j=0}^{\infty} \pi_j p_{ji} z^i$$

$$= \sum_{i=0}^{\infty} \sum_{j=0}^{\infty} \pi_j z^i e^{-\lambda} \sum_{n=0}^{i} \binom{j}{n} p^n q^{j-n} \frac{\lambda^{i-n}}{(i-n)!}$$

Since $\sum_{i=0}^{\infty} \sum_{n=0}^{i} = \sum_{n=0}^{\infty} \sum_{i=n}^{\infty}$ we have

$$P(z) = e^{-\lambda} \sum_{j=0}^{\infty} \pi_j \sum_{n=0}^{\infty} \binom{j}{n} p^n q^{j-n} \sum_{i=n}^{\infty} z^i \frac{\lambda^{i-n}}{(i-n)!}$$

$$= e^{-\lambda} e^{\lambda z} \sum_{j=0}^{\infty} \pi_j \sum_{n=0}^{\infty} \binom{j}{n} (pz)^n q^{j-n}$$

Since $\binom{j}{n} = 0$ for $n > j$, then

$$P(z) = e^{\lambda(z-1)} \sum_{j=0}^{\infty} \pi_j (pz + q)^j$$

Therefore

$$P(z) = e^{\lambda(z-1)} P[1 + p(z-1)]$$

(d) The expression is clearly true for $n = 1$ by part (c). By induction, assume it is true for arbitrary n, in which case

$$P(z) = e^{\lambda(z-1)(1+p+p^2+\cdots+p^{n-1})} P[1 + p^n(z-1)]$$

Now substitute $1 + p^n(z-1)$ for z in the result from part (c) to give

$$P[1 + p^n(z-1)] = e^{\lambda p^n(z-1)} P[1 + p^{n+1}(z-1)]$$

Thus

$$P(z) = e^{\lambda(z-1)(1+p+p^2+\cdots+p^n)} P[1 + p^{n+1}(z-1)]$$

which completes the induction.

(e) From (d),

$$P(z) = e^{\lambda(z-1)(1+p+p^2+\cdots+p^n)} P[1 + p^{n+1}(z-1)]$$

for all $n = 0, 1, 2, \ldots$. Consider this expression as $n \to \infty$. This gives

$$P(z) = e^{\lambda(z-1)\sum_{n=0}^{\infty} p^n} P(1) \qquad (p^{n+1} \to 0 \text{ since } 0 < p < 1)$$

or

$$P(z) = e^{\lambda(z-1)\frac{1}{1-p}} \qquad (P(1) = 1)$$

From Eq. (1.60) we recognize this as the z-transform of a Poisson distribution with parameter λ/q. Hence

$$\pi_i = \frac{\left(\dfrac{\lambda}{q}\right)^i e^{-\frac{\lambda}{q}}}{i!} \qquad \square$$

PROBLEM 2.8

Show that any point in or on the equilateral triangle of unit height represents a three-component probability vector in the sense that the sum of the distances from any such point to each of the three sides must always equal unity; that is, $p_1 + p_2 + p_3 = 1$.

SOLUTION

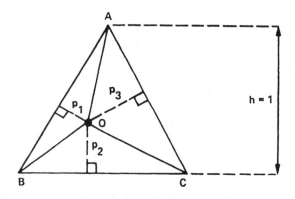

Let $\triangle ABC$ denote triangle ABC and let $|AB|$ denote the length of side AB. Then

$$\text{Area } \triangle ABC = \text{Area } \triangle AOB + \text{Area } \triangle BOC + \text{Area } \triangle AOC$$

Therefore

$$|BC|\frac{h}{2} = |AB|\frac{p_1}{2} + |BC|\frac{p_2}{2} + |AC|\frac{p_3}{2}$$

But $|BC| = |AB| = |AC|$ since $\triangle ABC$ is equilateral. So

$$h = p_1 + p_2 + p_3$$

Since $h = 1$ we have

$$p_1 + p_2 + p_3 = 1 \qquad \square$$

PROBLEM 2.9

Consider a "pure birth" process (i.e., a birth–death process where all death rates are zero). Moreover, assume that the process has constant birth rate λ. Let us consider an interval of length T, which we divide up into m segments each of length T/m. Define $\Delta t = T/m$ and $o(\Delta t)$ as any function that goes to zero faster than Δt.

(a) For Δt small, find the probability that a single arrival occurs in each of exactly k of the m intervals and that no arrivals occur in the remaining $m - k$ intervals.

(b) Consider the limit as $\Delta t \to 0$, that is, as $m \to \infty$ for fixed T, and evaluate the probability $P_k(T)$ that exactly k arrivals occur in the interval of length T.

SOLUTION

The pure birth equations are

$$P[1 \text{ arrival in } \Delta t] = \lambda \, \Delta t + o(\Delta t)$$

$$P[0 \text{ arrivals in } \Delta t] = 1 - \lambda \, \Delta t + o(\Delta t)$$

(a) We have

$$P \begin{bmatrix} 1 \text{ arrival in exactly } k \text{ of the} \\ m \text{ intervals and } 0 \text{ arrivals in the} \\ \text{remaining } m - k \text{ intervals} \end{bmatrix}$$

$$= \left\{ \begin{array}{l} \text{number of ways} \\ k \text{ cells may be} \\ \text{selected from } m \end{array} \right\} [\lambda \, \Delta t + o(\Delta t)]^k [1 - \lambda \, \Delta t + o(\Delta t)]^{m-k}$$

$$= \binom{m}{k} (\lambda \, \Delta t)^k (1 - \lambda \, \Delta t)^{m-k} + o(\Delta t)$$

(b) As $\Delta t \to 0$, since $\Delta t = T/m$, then $m \to \infty$ (and $o(\Delta t) \to 0$).

$$P_k(T) = \lim_{\Delta t \to 0} \left[\binom{m}{k} (\lambda \, \Delta t)^k (1 - \lambda \, \Delta t)^{m-k} + o(\Delta t) \right]$$

$$= \lim_{m \to \infty} \frac{m!}{k! \, (m - k)!} \left(\frac{\lambda T}{m} \right)^k \left(1 - \frac{\lambda T}{m} \right)^{m-k}$$

$$= \frac{(\lambda T)^k}{k!} \lim_{m \to \infty} \left(\frac{m}{m} \right) \left(\frac{m - 1}{m} \right) \cdots \left(\frac{m - k + 1}{m} \right) \left(1 - \frac{\lambda T}{m} \right)^{m-k}$$

$$= \frac{(\lambda T)^k}{k!} \lim_{m \to \infty} \left(1 - \frac{\lambda T}{m} \right)^{m-k} \quad (k \text{ fixed})$$

Therefore

$$P_k(T) = \frac{(\lambda T)^k}{k!} e^{-\lambda T} \qquad \qquad \square$$

PROBLEM 2.10

Consider a population of bacteria of size $N(t)$ at time t for which $N(0) = 1$. We consider this to be a pure birth process in which any member of the population will split into two new members in the interval $(t, t + \Delta t)$ with probability $\lambda \, \Delta t + o(\Delta t)$ or will remain unchanged in this interval with probability $1 - \lambda \, \Delta t + o(\Delta t)$ as $\Delta t \to 0$.

(a) Let $P_k(t) = P[N(t) = k]$ and write down the set of differential-difference equations that must be satisfied by these probabilities.

(b) From part (a) show that the z-transform $P(z, t)$ for $N(t)$ must satisfy

$$P(z, t) = \frac{ze^{-\lambda t}}{1 - z + ze^{-\lambda t}}$$

(c) Find $E[N(t)]$.

(d) Solve for $P_k(t)$.

(e) Solve for $P(z, t)$, $E[N(t)]$, and $P_k(t)$ that satisfy the initial condition $N(0) = n \geq 1$.

(f) Consider the corresponding deterministic problem in which each bacterium splits into two every $1/\lambda$ sec and compare with the answer in part (c).

SOLUTION

(a) The differential-difference equations (1.38) for this system are

$$P_k(t + \Delta t) = (1 - k\lambda \, \Delta t)P_k(t) + (k - 1)\lambda \, \Delta t \, P_{k-1}(t) + o(\Delta t)$$

$$P_k(t + \Delta t) - P_k(t) = (k - 1)\lambda \, \Delta t \, P_{k-1}(t) - k\lambda \, \Delta t \, P_k(t) + o(\Delta t)$$

$$\frac{dP_k(t)}{dt} = (k - 1)\lambda P_{k-1}(t) - k\lambda P_k(t) \qquad k \geq 1$$

$$\frac{dP_0(t)}{dt} = 0 \qquad \text{since } P_0(t) = 0 \text{ for all } t$$

$$(P_1(0) = 1 \text{ and pure birth})$$

(b) First we must form the z-transform of the above difference equations. Let $P(z, t) = \sum_{k=0}^{\infty} P_k(t)z^k$. We observe that

$$\frac{\partial P(z, t)}{\partial t} = \sum_{k=0}^{\infty} \frac{dP_k(t)}{dt} z^k$$

Using part (a), multiply the kth equation by z^k and sum to yield

$$\frac{\partial P(z, t)}{\partial t} = \sum_{k=1}^{\infty} \left[(k - 1)\lambda P_{k-1}(t)z^k - k\lambda P_k(t)z^k \right]$$

$$= \lambda z^2 \sum_{k=2}^{\infty}(k-1)P_{k-1}(t)z^{k-2} - \lambda z \sum_{k=1}^{\infty} kP_k(t)z^{k-1}$$

$$= \lambda z^2 \frac{\partial P(z,t)}{\partial z} - \lambda z \frac{\partial P(z,t)}{\partial z}$$

Thus

$$\frac{\partial P(z,t)}{\partial t} = \lambda z(z-1)\frac{\partial P(z,t)}{\partial z}$$

or

$$\frac{\partial P(z,t)}{\partial t} + \lambda z(1-z)\frac{\partial P(z,t)}{\partial z} = 0$$

Now we must solve this partial differential equation (p.d.e.). Anticipating the solution to Problem 2.14(b), let us solve the (more general) p.d.e.

$$\frac{\partial P(z,t)}{\partial t} + (\lambda z - \gamma)(1-z)\frac{\partial P(z,t)}{\partial z} = 0$$

(Note that the equation derived above is the special case when $\gamma = 0$.) We solve this linear first-order p.d.e. by the Lagrange method as described on pages 696–697 of [SYSK 60]. Thus we first must solve the subsidiary equation

$$\frac{dt}{1} = \frac{dz}{(\lambda z - \gamma)(1-z)}$$

which becomes, upon using partial fraction expansion,

$$\frac{dt}{1} = \frac{dz}{\lambda - \gamma}\left[\frac{\lambda}{\lambda z - \gamma} + \frac{1}{1-z}\right]$$

or

$$\frac{\lambda\, dz}{\lambda z - \gamma} - \frac{dz}{z-1} = (\lambda - \gamma)\, dt$$

Integrating yields

$$\log_e\left(\frac{\lambda z - \gamma}{z-1}\right) = (\lambda - \gamma)t + C$$

and thus

$$\frac{\lambda z - \gamma}{z-1} = Ke^{(\lambda-\gamma)t}$$

or

$$\frac{\gamma - \lambda z}{1-z}e^{-(\lambda-\gamma)t} = K$$

The general solution is now given as

$$P(z,t) = f\left[\frac{\gamma - \lambda z}{1 - z} e^{-(\lambda - \gamma)t}\right]$$

where the function f must be determined by the initial condition $P(z,0) = z$. This condition gives

$$f\left[\frac{\gamma - \lambda z}{1 - z}\right] = z$$

and so f satisfies

$$f(y) = \frac{y - \gamma}{y - \lambda}$$

Therefore

$$P(z,t) = \frac{\dfrac{\gamma - \lambda z}{1 - z} e^{-(\lambda - \gamma)t} - \gamma}{\dfrac{\gamma - \lambda z}{1 - z} e^{-(\lambda - \gamma)t} - \lambda}$$

$$= \frac{\gamma - \lambda z - \gamma(1 - z)\, e^{(\lambda - \gamma)t}}{\gamma - \lambda z - \lambda(1 - z)e^{(\lambda - \gamma)t}}$$

$$P(z,t) = \frac{\gamma(1 - e^{(\lambda - \gamma)t}) - (\lambda - \gamma e^{(\lambda - \gamma)t})z}{\gamma - \lambda e^{(\lambda - \gamma)t} - \lambda(1 - e^{(\lambda - \gamma)t})z}$$

which is the solution of the (more general) p.d.e. The special case $\gamma = 0$ gives the desired solution

$$P(z,t) = \frac{ze^{-\lambda t}}{1 - z + ze^{-\lambda t}}$$

(c) Differentiating gives

$$E[N(t)] = \left.\frac{\partial P(z,t)}{\partial z}\right|_{z=1} = \left.\frac{e^{-\lambda t}}{[1 - z + ze^{-\lambda t}]^2}\right|_{z=1}$$

$$E[N(t)] = e^{\lambda t}$$

(d) We have $P(z,t) = ze^{-\lambda t}/(1 - [1 - e^{-\lambda t}]z)$. To invert this z-transform we note from Entry 2, Table 1.3 that

$$e^{-\lambda t}[1 - e^{-\lambda t}]^k \Leftrightarrow \frac{e^{-\lambda t}}{1 - [1 - e^{-\lambda t}]z}$$

Using this result and Entry 4, Table 1.4 we have

$$P_k(t) = \begin{cases} 0 & k = 0 \text{ (pure birth)} \\ e^{-\lambda t}[1 - e^{-\lambda t}]^{k-1} & k \geq 1 \end{cases}$$

(e) When $N(0) = n \geq 1$, this is the same as having n independent birth processes each with $N(0) = 1$. The z-transform is the n-fold product of the z-transform derived in part (b); that is,

$$P(z,t) = \left(\frac{ze^{-\lambda t}}{1 - z + ze^{-\lambda t}} \right)^n$$

Thus, for the case $n \geq 1$, we have

$$E[N(t)] = ne^{\lambda t} \qquad \text{(sum of expectations)}$$

$$P(z,t) = \frac{e^{-n\lambda t} z^n}{\left(1 - [1 - e^{-\lambda t}]z \right)^n}$$

Inverting $P(z,t)$ we first find from Entry 5, Table 1:3 that

$$\frac{e^{-n\lambda t}}{\left(1 - [1 - e^{-\lambda t}]z \right)^n} \Leftrightarrow e^{-n\lambda t} \binom{k+n-1}{n-1} [1 - e^{-\lambda t}]^k$$

Now applying Entry 4, Table 1.4 we have

$$P_k(t) = \begin{cases} 0 & k < n \text{ (pure birth)} \\ \binom{k-1}{n-1} e^{-n\lambda t} [1 - e^{-\lambda t}]^{k-n} & k \geq n \end{cases}$$

(f) In the deterministic problem, each individual will split into 2 every $1/\lambda$ seconds exactly. Since $N(0) = 1$, and since every $1/\lambda$ seconds the population doubles, then

$$N(t) = 2^{\lambda t} \qquad t = \frac{k}{\lambda}, \, k = 0, 1, 2, \ldots$$

Note that this growth rate is less than $e^{\lambda t}$ for the random case. This is exactly analogous to the situation when \$1.00 is compounded annually at 100% interest to yield \$2^n$ in n years as opposed to \$e^n$ if compounded continuously. □

PROBLEM 2.11

Consider a birth–death process with coefficients

$$\lambda_k = \begin{cases} \lambda & k = 0 \\ 0 & k \neq 0 \end{cases} \qquad \mu_k = \begin{cases} \mu & k = 1 \\ 0 & k \neq 1 \end{cases}$$

which corresponds to an M/M/1 queueing system where there is no room for waiting customers.

(a) Give the differential-difference equations for $P_k(t)$ ($k = 0, 1$).

(b) Solve these equations and express the answers in terms of $P_0(0)$ and $P_1(0)$.

SOLUTION

(a) The differential-difference equations for this system are

$$\frac{dP_0(t)}{dt} = -\lambda P_0(t) + \mu P_1(t)$$

$$\frac{dP_1(t)}{dt} = -\mu P_1(t) + \lambda P_0(t)$$

(b) Since $P_0(t) + P_1(t) = 1$, we may rewrite the first equation from part (a) as

$$\frac{dP_0(t)}{dt} + (\lambda + \mu)P_0(t) = \mu$$

From the theory of ordinary linear differential equations [KLEI 75], we recognize that the homogeneous solution must be of the form $Ae^{-(\lambda+\mu)t}$; we find that the particular solution is a constant (say, B) and so

$$P_0(t) = B + Ae^{-(\lambda+\mu)t}$$

Substituting into our differential equation, we have $B = \mu/(\lambda + \mu)$. Evaluating the solution at $t = 0$, we have $A = P_0(0) - \mu/(\lambda + \mu)$. Hence

$$P_0(t) = \frac{\mu}{\lambda + \mu} + \left(P_0(0) - \frac{\mu}{\lambda + \mu}\right)e^{-(\lambda+\mu)t}$$

$P_1(t)$ may be found from $P_1(t) = 1 - P_0(t)$ or by symmetry of the defining equations from part (a) to yield

$$P_1(t) = \frac{\lambda}{\lambda + \mu} + \left(P_1(0) - \frac{\lambda}{\lambda + \mu}\right)e^{-(\lambda+\mu)t} \qquad \square$$

PROBLEM 2.12

Consider a birth–death queueing system in which

$$\lambda_k = \lambda \qquad k \geq 0$$

$$\mu_k = k\mu \qquad k \geq 0$$

(a) For all k, find the differential-difference equations for

$$P_k(t) = P[k \text{ in system at time } t]$$

(b) Define the z-transform

$$P(z,t) = \sum_{k=0}^{\infty} P_k(t)z^k$$

and find the *partial* differential equation that $P(z,t)$ must satisfy.

(c) Show that the solution to this equation is

$$P(z,t) = \exp\left[\frac{\lambda}{\mu}(1 - e^{-\mu t})(z - 1)\right]$$

with the initial condition $P_0(0) = 1$.

(d) Comparing the solution in part (c) with Eq. (1.60), give the expression for $P_k(t)$ by inspection.

(e) Find the limiting values for these probabilities as $t \to \infty$.

SOLUTION

(a) The differential-difference equations for this system are

$$\frac{dP_k(t)}{dt} = -(\lambda + k\mu)P_k(t) + \lambda P_{k-1}(t) + (k + 1)\mu P_{k+1}(t) \qquad k \geq 1$$

$$\frac{dP_0(t)}{dt} = -\lambda P_0(t) + \mu P_1(t) \qquad k = 0$$

(b) For $k \geq 1$, we multiply the kth equation by z^k, sum on k, and then add the equation for $k = 0$ to get

$$\sum_{k=0}^{\infty} \frac{dP_k(t)}{dt}z^k = -\lambda \sum_{k=0}^{\infty} P_k(t)z^k - \mu z \sum_{k=1}^{\infty} kP_k(t)z^{k-1}$$

$$+ \lambda z \sum_{k=1}^{\infty} P_{k-1}(t)z^{k-1} + \mu \sum_{k=0}^{\infty}(k + 1)P_{k+1}(t)z^k$$

$$\frac{\partial P(z,t)}{\partial t} = -\lambda P(z,t) - \mu z\frac{\partial P(z,t)}{\partial z} + \lambda z P(z,t) + \mu\frac{\partial P(z,t)}{\partial z}$$

$$\frac{\partial P(z,t)}{\partial t} = \lambda(z - 1)P(z,t) - \mu(z - 1)\frac{\partial P(z,t)}{\partial z}$$

(c) The partial differential equation derived in part (b) may be solved by the method used in Problem 2.10(b) (see the reference mentioned there). Writing our equation in the form

$$\frac{\partial P(z,t)}{\partial t} + \mu(z - 1)\frac{\partial P(z,t)}{\partial z} = \lambda(z - 1)P(z,t)$$

we see that the following subsidiary equations must first be solved

$$\frac{dt}{1} = \frac{dz}{\mu(z - 1)} = \frac{dP(z,t)}{\lambda(z - 1)P(z,t)}$$

The first equation becomes

$$\frac{dz}{z - 1} = \mu\,dt$$

which yields, upon integration,

$$\log_e(z - 1) = \mu t + C_1$$

and thus

$$z - 1 = K_1 e^{\mu t}$$

or

$$(z - 1)e^{-\mu t} = K_1$$

The second equation becomes

$$\frac{dP(z, t)}{P(z, t)} = \frac{\lambda}{\mu} dz$$

which yields

$$\log_e P(z, t) = \frac{\lambda}{\mu} z + C_2$$

and thus

$$P(z, t) = K_2 e^{\frac{\lambda}{\mu} z}$$

or

$$P(z, t)e^{-\frac{\lambda}{\mu} z} = K_2$$

The general solution is now given as

$$P(z, t)e^{-\frac{\lambda}{\mu} z} = f[(z - 1)e^{-\mu t}]$$

where the function f must be determined by the initial condition $P_0(0) = 1$. This condition [which is equivalent to $P(z, 0) = 1$] gives

$$e^{-\frac{\lambda}{\mu} z} = f(z - 1)$$

and so f satisfies

$$f(y) = e^{-\frac{\lambda}{\mu}(y + 1)}$$

Therefore

$$P(z, t) = f[(z - 1)e^{-\mu t}]e^{\frac{\lambda}{\mu} z} = e^{-\frac{\lambda}{\mu}[(z-1)e^{-\mu t} + 1]}e^{\frac{\lambda}{\mu} z}$$

which gives the desired solution

$$P(z, t) = \exp\left[\frac{\lambda}{\mu}(1 - e^{-\mu t})(z - 1)\right]$$

(The interested reader may wish to substitute this solution into the p.d.e. above.)

(d) $P_k(t)$ is clearly Poisson with parameter $(\lambda/\mu)[1 - e^{-\mu t}]$; that is,

$$P_k(t) = \frac{\left(\dfrac{\lambda}{\mu}\right)^k [1 - e^{-\mu t}]^k}{k!} e^{-\frac{\lambda}{\mu}[1 - e^{-\mu t}]}$$

(e) As $t \to \infty$, $e^{-\mu t} \to 0$ and $P_k(t) \to p_k$.

$$p_k = \frac{\left(\dfrac{\lambda}{\mu}\right)^k}{k!} e^{-\frac{\lambda}{\mu}} \qquad \left[\text{Poisson with parameter } \frac{\lambda}{\mu}\right] \qquad \square$$

PROBLEM 2.13

Consider a system in which the birth rate decreases and the death rate increases as the number in the system k increases; that is,

$$\lambda_k = \begin{cases} (K - k)\lambda & k \le K \\ 0 & k \ge K \end{cases} \qquad \mu_k = \begin{cases} k\mu & k \le K \\ 0 & k \ge K \end{cases}$$

Write down the differential-difference equations for

$$P_k(t) = P[k \text{ in system at time } t]$$

SOLUTION

The differential-difference equations are

$$\frac{dP_0(t)}{dt} = \mu P_1(t) - K\lambda P_0(t) \qquad k = 0$$

$$\frac{dP_k(t)}{dt} = \lambda(K - k + 1)P_{k-1}(t) + (k + 1)\mu P_{k+1}(t)$$

$$\qquad - [(K - k)\lambda + k\mu]P_k(t) \qquad k = 1, 2, \ldots, K - 1$$

$$\frac{d P_K(t)}{dt} = \lambda P_{K-1}(t) - K\mu P_K(t) \qquad k = K \qquad \square$$

PROBLEM 2.14

Consider the case of a linear birth–death process in which $\lambda_k = k\lambda$ and $\mu_k = k\mu$.

(a) Find the partial differential equation that is satisfied by $P(z, t) = \sum_{k=0}^{\infty} P_k(t)z^k$.

(b) Assuming that the population size is one at time zero, show that the function that satisfies the equation in part (a) is

$$P(z, t) = \frac{\mu(1 - e^{(\lambda - \mu)t}) - (\lambda - \mu e^{(\lambda - \mu)t})z}{\mu - \lambda e^{(\lambda - \mu)t} - \lambda(1 - e^{(\lambda - \mu)t})z}$$

(c) Expanding $P(z, t)$ in a power series show that

$$P_k(t) = [1 - \alpha(t)][1 - \beta(t)][\beta(t)]^{k-1} \qquad k = 1, 2, \ldots$$

$$P_0(t) = \alpha(t)$$

and find $\alpha(t)$ and $\beta(t)$.

(d) Find the mean and variance for the number in system at time t.

(e) Find the limiting probability that the population dies out by time t for $t \to \infty$.

SOLUTION

(a) The differential-difference equations are

$$\frac{dP_k(t)}{dt} = -(k\lambda + k\mu)P_k(t) + (k - 1)\lambda P_{k-1}(t) + (k + 1)\mu P_{k+1}(t) \qquad k \geq 1$$

$$\frac{dP_0(t)}{dt} = \mu P_1(t) \qquad k = 0$$

Multiplying the kth equation by z^k and summing we obtain

$$\sum_{k=0}^{\infty} \frac{dP_k(t)}{dt} z^k = -(\lambda + \mu) \sum_{k=1}^{\infty} kP_k(t)z^k + \lambda \sum_{k=1}^{\infty} (k - 1)P_{k-1}(t)z^k$$

$$+ \mu \sum_{k=0}^{\infty} (k + 1)P_{k+1}(t)z^k$$

$$\frac{\partial P(z, t)}{\partial t} = -(\lambda + \mu)z \frac{\partial P(z, t)}{\partial z} + \lambda z^2 \frac{\partial P(z, t)}{\partial z} + \mu \frac{\partial P(z, t)}{\partial z}$$

$$\frac{\partial P(z, t)}{\partial t} = (\lambda z - \mu)(z - 1) \frac{\partial P(z, t)}{\partial z}$$

(b) As part of the solution to Problem 2.10(b), the equation

$$\frac{\partial P(z, t)}{\partial t} + (\lambda z - \gamma)(1 - z) \frac{\partial P(z, t)}{\partial z} = 0$$

was solved. We see that the partial differential equation derived in part (a) above is the same as this equation with μ in place of γ. Thus the solution obtained in Problem 2.10(b) may be used, and we find that

$$P(z, t) = \frac{\mu(1 - e^{(\lambda - \mu)t}) - (\lambda - \mu e^{(\lambda - \mu)t})z}{\mu - \lambda e^{(\lambda - \mu)t} - \lambda(1 - e^{(\lambda - \mu)t})z}$$

as desired.

(c) Set

$$a = a(t) = \mu(1 - e^{(\lambda - \mu)t}) \qquad b = b(t) = \lambda - \mu e^{(\lambda - \mu)t}$$

$$c = c(t) = \mu - \lambda e^{(\lambda - \mu)t} \qquad d = d(t) = \lambda(1 - e^{(\lambda - \mu)t})$$

Then

$$P(z,t) = \frac{a - bz}{c - dz} = \frac{b}{d} + \frac{ad - bc}{cd\left(1 - \frac{d}{c}z\right)}$$

$$P(z,t) = \frac{b}{d} + \frac{ad - bc}{cd} \sum_{k=0}^{\infty} \left(\frac{d}{c}z\right)^k$$

$$k = 0: \quad P_0(t) = P(0,t) = \frac{b}{d} + \frac{ad - bc}{cd} = \frac{a}{c}$$

$$P_0(t) = \frac{a}{c} = \frac{\mu(1 - e^{(\lambda - \mu)t})}{\mu - \lambda e^{(\lambda - \mu)t}} \triangleq \alpha(t)$$

$$k \geq 1: \quad P_k(t) = \frac{ad - bc}{cd}\left(\frac{d}{c}\right)^k = \frac{ad - bc}{cd}\left(\frac{d}{c}\right)\left(\frac{d}{c}\right)^{k-1}$$

$$P_k(t) = \frac{ad - bc}{c^2}\left(\frac{d}{c}\right)^{k-1}$$

We take advantage of the fact that $a - b = c - d = \mu - \lambda$. Thus

$$ad - bc = ad + (c - a - d)c = (c - a)(c - d)$$

and so

$$P_k(t) = \frac{(c - a)(c - d)}{c^2}\left(\frac{d}{c}\right)^{k-1} = \left(1 - \frac{a}{c}\right)\left(1 - \frac{d}{c}\right)\left(\frac{d}{c}\right)^{k-1}$$

Therefore

$$P_k(t) = [1 - \alpha(t)][1 - \beta(t)][\beta(t)]^{k-1}$$

where

$$\alpha(t) \triangleq \frac{a}{c} = \frac{\mu(1 - e^{(\lambda - \mu)t})}{\mu - \lambda e^{(\lambda - \mu)t}}$$

and

$$\beta(t) \triangleq \frac{d}{c} = \frac{\lambda(1 - e^{(\lambda - \mu)t})}{\mu - \lambda e^{(\lambda - \mu)t}}$$

(d) Generalizing Eq. (1.24), we see that

$$\overline{N}(t) = \frac{\partial P(z,t)}{\partial z}\bigg|_{z=1}$$

Recall that

$$P(z,t) = \frac{a - bz}{c - dz}$$

and so

$$\frac{\partial P(z,t)}{\partial z} = \frac{ad - bc}{(c - dz)^2} = \frac{(c - a)(c - d)}{(c - dz)^2}$$

Evaluating this partial derivative at $z = 1$, we find

$$\overline{N}(t) = e^{(\lambda - \mu)t}$$

Now

$$\sigma^2_{N(t)} = \overline{N^2}(t) - \left(\overline{N}(t)\right)^2$$

$$= \sum_{k=1}^{\infty} k(k - 1)P_k(t) + \overline{N}(t) - \left(\overline{N}(t)\right)^2$$

But, generalizing Eq. (1.25), we have

$$\sum_{k=1}^{\infty} k(k - 1)P_k(t) = \left.\frac{\partial^2 P(z,t)}{\partial z^2}\right|_{z=1} = \left.\frac{2d(c - a)(c - d)}{(c - dz)^3}\right|_{z=1}$$

and so

$$\sum_{k=1}^{\infty} k(k - 1)P_k(t) = \frac{2\lambda e^{(\lambda - \mu)t}}{\mu - \lambda}(1 - e^{(\lambda - \mu)t})$$

Therefore

$$\sigma^2_{N(t)} = \frac{\mu + \lambda}{\mu - \lambda}e^{(\lambda - \mu)t}(1 - e^{(\lambda - \mu)t})$$

(e) We have

$$p_0 = \lim_{t \to \infty} P_0(t) = \lim_{t \to \infty} \frac{\mu(1 - e^{(\lambda - \mu)t})}{\mu - \lambda e^{(\lambda - \mu)t}}$$

$$p_0 = \begin{cases} 1 & \text{if } \lambda \le \mu \\ \dfrac{\mu}{\lambda} & \text{if } \lambda > \mu \end{cases}$$ □

PROBLEM 2.15

Consider a linear birth–death process for which $\lambda_k = k\lambda + \alpha$ and $\mu_k = k\mu$.

(a) Find the differential-difference equations that must be satisfied by $P_k(t)$.
(b) From (a) find the partial-differential equation that must be satisfied by the time-dependent transform defined as

$$P(z,t) = \sum_{k=0}^{\infty} P_k(t)z^k$$

(c) What is the value of $P(1, t)$?

(d) Assuming that the population size begins with i members at time 0, find an ordinary differential equation for $\overline{N}(t)$ and then solve for $\overline{N}(t)$. Consider the case $\lambda = \mu$ as well as $\lambda \neq \mu$.

(e) Find the limiting value for $\overline{N}(t)$ in the case $\lambda < \mu$ (as $t \to \infty$).

SOLUTION

(a) The differential-difference equations are

$$\frac{dP_k(t)}{dt} = -[k(\mu + \lambda) + \alpha]P_k(t) + [(k-1)\lambda + \alpha]P_{k-1}(t)$$

$$+ (k+1)\mu P_{k+1}(t) \qquad k \geq 1$$

$$\frac{dP_0(t)}{dt} = \mu P_1(t) - \alpha P_0(t) \qquad k = 0$$

(b) Multiplying the kth equation by z^k and summing yields

$$\sum_{k=0}^{\infty} \frac{dP_k(t)}{dt} z^k = -(\lambda + \mu) \sum_{k=1}^{\infty} k P_k(t) z^k - \alpha \sum_{k=0}^{\infty} P_k(t) z^k$$

$$+ \lambda \sum_{k=1}^{\infty} (k-1) P_{k-1}(t) z^k + \alpha \sum_{k=1}^{\infty} P_{k-1}(t) z^k$$

$$+ \mu \sum_{k=0}^{\infty} (k+1) P_{k+1}(t) z^k$$

$$\frac{\partial P(z, t)}{\partial t} = -(\lambda + \mu) z \frac{\partial P(z, t)}{\partial z} - \alpha P(z, t) + \lambda z^2 \frac{\partial P(z, t)}{\partial z}$$

$$+ \alpha z P(z, t) + \mu \frac{\partial P(z, t)}{\partial z}$$

$$\frac{\partial P(z, t)}{\partial t} = (z - 1) \left[(\lambda z - \mu) \frac{\partial P(z, t)}{\partial z} + \alpha P(z, t) \right]$$

(c) Differentiating gives

$$P(1, t) = \sum_{k=0}^{\infty} P_k(t) z^k \bigg|_{z=1} = 1 \qquad \text{for all } t$$

The average number in system at time t, $\overline{N}(t)$, is given by

$$\overline{N}(t) = \lim_{z \to 1} \frac{\partial P(z, t)}{\partial z} = \lim_{z \to 1} \frac{\partial}{\partial z} \sum_{k=0}^{\infty} P_k(t) z^k$$

$$= \lim_{z \to 1} \sum_{k=1}^{\infty} k P_k(t) z^{k-1} = \sum_{k=1}^{\infty} k P_k(t)$$

(d) We must form the limit as $z \to 1$ in the partial differential equation developed in part (b); this will give us the required equation for $\bar{N}(t)$. To this end, we must evaluate

$$\lim_{z \to 1} \frac{\partial P(z,t)}{\partial t} \cdot \frac{1}{z-1} = \lim_{z \to 1} \frac{\partial^2 P(z,t)}{\partial z \, \partial t} \cdot 1 \qquad \text{(by L'Hospital)}$$

$$= \frac{d}{dt} \lim_{z \to 1} \frac{\partial P(z,t)}{\partial z} = \frac{d}{dt} \bar{N}(t)$$

and also

$$\lim_{z \to 1} \left[(\lambda z - \mu) \frac{\partial P(z,t)}{\partial z} + \alpha P(z,t) \right] = (\lambda - \mu)\bar{N}(t) + \alpha$$

Thus

$$\frac{d}{dt} \bar{N}(t) = (\lambda - \mu)\bar{N}(t) + \alpha$$

(i) $\lambda = \mu$

$$\frac{d}{dt} \bar{N}(t) = \alpha \text{ and thus}$$

$$\bar{N}(t) = \alpha t + \bar{N}(0) = \alpha t + i$$

(ii) $\lambda \neq \mu$

$$\frac{d}{dt} \bar{N}(t) - (\lambda - \mu)\bar{N}(t) = \alpha. \text{ Solving this differential equation as in part}$$

(b) of Problem 2.11, we obtain

$$\bar{N}(t) = \frac{\alpha}{\mu - \lambda} + \left(i - \frac{\alpha}{\mu - \lambda} \right) e^{(\lambda - \mu)t}$$

(e) For the case $\lambda < \mu$, as $t \to \infty$ then $e^{(\lambda - \mu)t} \to 0$. Hence

$$\lim_{t \to \infty} \bar{N}(t) = \frac{\alpha}{\mu - \lambda} \qquad \text{for } \lambda < \mu \qquad \square$$

PROBLEM 2.16

Consider Eq. (1.62) with $\mu_k = 0$ (i.e., pure birth) and define the Laplace transform

$$P_k^*(s) = \int_0^\infty P_k(t) e^{-st} \, dt$$

For our initial condition assume $P_0(t) = 1$ for $t = 0$. Transform this pure birth version of Eq. (1.62) to obtain a set of linear difference equations in $\{P_k^*(s)\}$.

(a) Show that the solution to the set of equations is

$$P_k^*(s) = \frac{\prod_{i=0}^{k-1} \lambda_i}{\prod_{i=0}^{k}(s + \lambda_i)}$$

(b) From (a) find $P_k(t)$ for the case $\lambda_i = \lambda$ $(i = 0, 1, 2, \ldots)$.

SOLUTION

Transforming the equations of motion, we get

$$sP_k^*(s) - P_k(0) = -\lambda_k P_k^*(s) + \lambda_{k-1} P_{k-1}^*(s) \qquad k \geq 1$$

and

$$sP_0^*(s) - P_0(0) = -\lambda_0 P_0^*(s) \qquad k = 0$$

The initial condition $P_0(0) = 1$ implies $P_k(0) = 0$ for $k \geq 1$. Thus

$$(s + \lambda_k)P_k^*(s) = \lambda_{k-1} P_{k-1}^*(s) \qquad k \geq 1$$

and

$$(s + \lambda_0)P_0^*(s) = 1 \qquad k = 0$$

(a) We have $P_0^*(s) = 1/(s + \lambda_0)$. Solving recursively

$$P_1^*(s) = \frac{\lambda_0}{(s + \lambda_0)(s + \lambda_1)}$$

In general, we have that

$$P_k^*(s) = \frac{\prod_{i=0}^{k-1} \lambda_i}{\prod_{i=0}^{k}(s + \lambda_i)} \qquad k \geq 0$$

(b) For the case $\lambda_i = \lambda$ $(i = 0, 1, 2, \ldots)$ from part (a),

$$P_k^*(s) = \frac{\lambda^k}{(s + \lambda)^{k+1}} \qquad k \geq 0$$

From Entry 4, Table 1.1 we have that

$$\frac{t^k}{k!}e^{-\lambda t} \Leftrightarrow \frac{1}{(s + \lambda)^{k+1}}$$

and so

$$P_k^*(s) = \frac{\lambda^k}{(s + \lambda)^{k+1}} \Leftrightarrow \frac{(\lambda t)^k}{k!}e^{-\lambda t}$$

Hence, for $t \geq 0$,

$$P_k(t) = \frac{(\lambda t)^k}{k!}e^{-\lambda t} \qquad \square$$

PROBLEM 2.17

Consider a time interval $(0, t)$ during which a Poisson process generates arrivals at an average rate λ. Derive the k-stage Erlangian pdf given in Eq. (1.27). Carry this out by considering two events: the event that exactly $k - 1$ arrivals occur in the interval $(0, t - \Delta t)$ and the event that exactly one arrival occurs in the interval $(t - \Delta t, t)$. Considering the limit as $\Delta t \to 0$ we immediately arrive at our desired result.

SOLUTION

Recall that the random variable of interest in Eq. (1.27) is X, the time interval required to collect k arrivals from a Poisson process, with $f_X(t)$ its pdf. To find $f_X(t)$, we first note that $f_X(t) \Delta t$ is the probability that k arrivals occur in $(0, t)$ with the kth arrival in $(t - \Delta t, t)$. Thus $f_X(t) \Delta t$ is simply the probability that there are $k - 1$ arrivals in $(0, t - \Delta t)$ and one arrival in $(t - \Delta t, t)$. Therefore we have

$$f_X(t)\,\Delta t = \left[\frac{[\lambda(t - \Delta t)]^{k-1}}{(k - 1)!} e^{-\lambda(t - \Delta t)} \right] \left[\lambda\,\Delta t\, e^{-\lambda\,\Delta t} \right]$$

$$= \frac{\lambda[\lambda(t - \Delta t)]^{k-1}}{(k - 1)!} e^{-\lambda t}\,\Delta t$$

Hence, as $\Delta t \to 0$, we obtain

$$f_X(t) = \frac{\lambda(\lambda t)^{k-1}}{(k - 1)!} e^{-\lambda t} \qquad \square$$

PROBLEM 2.18

A barber opens up for business at $t = 0$. Customers arrive at random in a Poisson fashion; that is, the pdf of interarrival time is $a(t) = \lambda e^{-\lambda t}$. Each haircut takes X sec (where X is some random variable). Find the probability P that the second arriving customer will not have to wait and also find W, the average value of his waiting time for the following two cases:

(i) $X = c = $ constant.
(ii) X is exponentially distributed with pdf:

$$b(x) = \mu e^{-\mu x}$$

SOLUTION

(i) $X = c = $ constant

The probability the second arriving customer will not have to wait is the probability that his interarrival time is $\geq c$.

$$P[\text{no waiting}] = \int_c^\infty \lambda e^{-\lambda t}\, dt = e^{-\lambda c}$$

The second customer will wait $c - y$ seconds if his interarrival time $y \leq c$. Otherwise he does not wait. Hence his average waiting time is

$$W = \int_0^c (c - y)P[y < \text{interarrival time} \leq y + dy] = \int_0^c (c - y)\lambda e^{-\lambda y}\, dy$$

$$W = c - \frac{1}{\lambda}[1 - e^{-\lambda c}]$$

(ii) X is exponentially distributed with pdf $b(x) = \mu e^{-\mu x}$

As in (i) above, the second customer does not have to wait iff his interarrival time is $\geq$ the first customer's service time.

$$P[\text{no waiting}] = P[\text{interarrival time} \geq \text{service time}]$$

$$= \int_0^\infty P[\text{interarrival time} \geq \text{service time} \,|$$

$$\text{service time} = x] \cdot \mu e^{-\mu x}\, dx$$

$$= \int_0^\infty e^{-\lambda x} \mu e^{-\mu x}\, dx = \frac{\mu}{\lambda + \mu}$$

The average waiting time given that the service time was x seconds is $x - (1/\lambda)[1 - e^{-\lambda x}]$ from case (i). Unconditioning on the service time gives

$$W = \int_0^\infty \left(x - \frac{1}{\lambda}[1 - e^{-\lambda x}] \right) \mu e^{-\mu x}\, dx$$

$$W = \frac{\lambda}{\mu(\mu + \lambda)} \qquad\qquad \square$$

PROBLEM 2.19

At $t = 0$ customer A places a request for service and finds all m servers busy and n other customers waiting for service in an M/M/m queueing system (see Section 1.5). All customers wait as long as necessary for service, waiting customers are served in order of arrival, and no new requests for service are permitted after $t = 0$. Service times are assumed to be mutually independent, identical, exponentially distributed random variables, each with mean duration $1/\mu$.

(a) Find the expected length of time customer A spends waiting for service in the queue.

(b) Find the expected length of time from the arrival of customer A at $t = 0$ until the system becomes completely empty (all customers complete service).

(c) Let X be the order of completion of service of customer A; that is, $X = k$ if A is the kth customer to complete service after $t = 0$. Find $P[X = k]$ $(k = 1, 2, \ldots, m + n + 1)$.

(d) Find the probability that customer A completes service before the customer immediately ahead of him in the queue.

(e) Let $\tilde{w}$ be the amount of time customer A waits for service. Find $P[\tilde{w} > x]$.

SOLUTION

(a) When all m servers are busy, an interdeparture time has mean $1/m\mu$. Customer A must wait in the queue until $n + 1$ such departures occur. Hence

$$E[\text{waiting time for A}] = \frac{n + 1}{m\mu}$$

(b) Adding the expected waiting time for customer A and the expected time to serve all remaining customers, we obtain

$$E[\text{time to empty system}] = \frac{n + 1}{m\mu} + \frac{1}{m\mu} + \cdots + \frac{1}{\mu}$$

$$= \frac{n + 1}{m\mu} + \frac{1}{\mu}\sum_{k=1}^{m}\frac{1}{k}$$

(c) Since $n + 1$ customers must leave before A enters service, we have

$$P[X = k] = 0 \qquad \text{for } k = 1, 2, \ldots, n, n + 1$$

For $n + 2 \leq k \leq m + n + 1$, A's departure order is equally likely to be any of the m possible values (by symmetry and the memoryless property). Thus

$$P[X = k] = \frac{1}{m} \qquad \text{for } k = n + 2, \ldots, m + n + 1$$

(d) Let B be the customer immediately ahead of A in the queue. A completes service before B if A enters service before B finishes (which occurs with probability $[m - 1]/m$) and, once in service, A finishes before B (which occurs with probability $\frac{1}{2}$). Hence

$$P[\text{A completes service before B}] = \frac{m - 1}{2m}$$

(e) $\tilde{w} = \tilde{d}_1 + \cdots + \tilde{d}_{n+1}$, where $\tilde{d}_i$ are interdeparture times, which are independent and identically distributed from an exponential distribution with mean $1/m\mu$. Therefore $\tilde{w}$ is a random variable with an $(n + 1)$-stage Erlangian pdf with parameter $m\mu$. That is, $\tilde{w}$ has pdf

$$f_{\tilde{w}}(x) = \frac{m\mu(m\mu x)^n}{n!}e^{-m\mu x}$$

Integrating (by parts repeatedly) we get

$$P[\tilde{w} > x] = \sum_{i=0}^{n}\frac{(m\mu x)^i}{i!}e^{-m\mu x} \qquad \square$$

PROBLEM 2.20

In this problem we wish to proceed from Eq. (1.76) to the transient solution in Eq. (1.77). Since $P^*(z, s)$ must converge in the region $|z| \le 1$ for $\text{Re}(s) > 0$, then, in this region, the zeros of the denominator in Eq. (1.76) must also be zeros of the numerator.

(a) Find those two values of z that give the denominator zeros, and denote them by $\alpha_1(s)$, $\alpha_2(s)$, where $|\alpha_2(s)| < |\alpha_1(s)|$.

(b) Using Rouché's theorem (see [KLEI 75]) show that the denominator of $P^*(z, s)$ has a single zero within the unit disk $|z| \le 1$.

(c) Requiring that the numerator of $P^*(z, s)$ vanish at $z = \alpha_2(s)$ from our earlier considerations, find an explicit expression for $P_0^*(s)$.

(d) Write $P^*(z, s)$ in terms of $\alpha_1(s) = \alpha_1$ and $\alpha_2(s) = \alpha_2$. Then show that this equation may be reduced to

$$P^*(z, s) = \frac{(z^i + \alpha_2 z^{i-1} + \cdots + \alpha_2^i) + \alpha_2^{i+1}/(1 - \alpha_2)}{\lambda \alpha_1 (1 - z/\alpha_1)}$$

(e) For $k \ge i$, using the fact that $|\alpha_2| < 1$ and that $\alpha_1 \alpha_2 = \mu/\lambda$ show that the inversion on z yields the following expression for $P_k^*(s)$, which is the Laplace transform for our transient probabilities $P_k(t)$:

$$P_k^*(s) = \frac{1}{\lambda} \left[\alpha_1^{i-k-1} + \left(\frac{\mu}{\lambda}\right) \alpha_1^{i-k-3} + \left(\frac{\mu}{\lambda}\right)^2 \alpha_1^{i-k-5} + \cdots \right.$$
$$\left. + \left(\frac{\mu}{\lambda}\right)^i \alpha_1^{-i-k-1} + \left(\frac{\lambda}{\mu}\right)^{k+1} \sum_{j=k+i+2}^{\infty} \left(\frac{\mu}{\lambda \alpha_1}\right)^j \right]$$

(f) In what follows we take advantage of Entry 4, Table 1.2, and also we make use of the following transform pair:

$$k \rho^{k/2} t^{-1} I_k(at) \Leftrightarrow \left[\frac{s + \sqrt{s^2 - 4\lambda\mu}}{2\lambda} \right]^{-k}$$

where $\rho = \lambda/\mu$, $a = 2\mu\sqrt{\rho}$, and $I_k(x)$ is the modified Bessel function of the first kind of order k as defined in Eq. (1.78). Using these facts and the simple relations among Bessel functions, namely,

$$\frac{2k}{x} I_k(x) = I_{k-1}(x) - I_{k+1}(x) \text{ and } I_k(x) = I_{-k}(x)$$

show that Eq. (1.77) is the inverse transform for the expression shown in part (e), thus establishing the transient solution for $k \ge i$.

(g) Starting with the equation in part (d), extend the applicability of Eq. (1.77) to the range $k < i$.

SOLUTION

(a) Equation (1.76) is

$$P^*(z, s) = \frac{z^{i+1} - \mu(1 - z)P_0^*(s)}{sz - (1 - z)(\mu - \lambda z)} = \frac{N(z, s)}{D(z, s)} = \frac{N}{D}$$

We may write $D = -[\lambda z^2 - (s + \mu + \lambda)z + \mu]$. The two zeros of D are

$$\alpha_1(s) = \frac{s + \mu + \lambda + \sqrt{(s + \mu + \lambda)^2 - 4\lambda\mu}}{2\lambda}$$

and

$$\alpha_2(s) = \frac{s + \mu + \lambda - \sqrt{(s + \mu + \lambda)^2 - 4\lambda\mu}}{2\lambda}$$

Thus $D = -\lambda[z - \alpha_1(s)][z - \alpha_2(s)]$. We now wish to show that $|\alpha_2(s)| < |\alpha_1(s)|$ for $\text{Re}(s) > 0$, or that $|\alpha_1(s)|^2 - |\alpha_2(s)|^2 > 0$. Thus, defining $h \overset{\Delta}{=} \sqrt{(s + \mu + \lambda)^2 - 4\lambda\mu}$ and substituting for $\alpha_1(s)$ and $\alpha_2(s)$ in the latter inequality yields the equivalent condition [which we denote by (#)]

$$(\#) \quad \text{Re}(s + \mu + \lambda)\text{Re}(h) + \text{Im}(s + \mu + \lambda)\text{Im}(h) > 0$$

Thus we need only show (#) for $\text{Re}(s) > 0$. There are three cases to consider.

Case (1): $\text{Im}(s) = 0$

We may express h as

$$h = \sqrt{s^2 + 2s(\mu + \lambda) + (\mu + \lambda)^2 - 4\lambda\mu} = \sqrt{s^2 + 2s(\mu + \lambda) + (\mu - \lambda)^2}$$

However, in this case we note that s is a positive real number. Therefore h is also real and positive, and so (#) holds.

Case (2): $\text{Im}(s) > 0$

In this case $\text{Re}(s + \mu + \lambda) > 0$ and $\text{Im}(s + \mu + \lambda) > 0$. Thus $s + \mu + \lambda$ may be represented as a point in the first quadrant of the complex plane. The complex number $(s + \mu + \lambda)^2$ must therefore be in the first or second quadrant. Subtracting a real number to form $(s + \mu + \lambda)^2 - 4\lambda\mu = h^2$ leaves us in the first or second quadrant, and taking the square root yields a point (h) in the first quadrant. That is. $\text{Re}(h) > 0$ and $\text{Im}(h) > 0$. Thus we see that (#) is satisfied.

Case (3): $\text{Im}(s) < 0$

In this case $\text{Re}(s + \mu + \lambda) > 0$ and $\text{Im}(s + \mu + \lambda) < 0$. In a manner similar to that used in Case (2) above, we find that $\text{Re}(h) > 0$ and $\text{Im}(h) < 0$. Thus once again (#) holds.

Since (#) is satisfied in all three cases, we have shown that, for $\text{Re}(s) > 0$, $|\alpha_2(s)| < |\alpha_1(s)|$ as desired.

(b) Decompose D as follows: $D = f_s(z) + g(z)$, where $f_s(z) = (s + \mu + \lambda)z$, $g(z) = -(\lambda z^2 + \mu)$. Then for $|z| = 1$ (and $\text{Re}(s) > 0$)

$$|f_s(z)| = |\lambda + \mu + s| \geq \lambda + \mu + \text{Re}(s) > \lambda + \mu$$

and

$$|g(z)| \leq \lambda + \mu$$

Thus, on $|z| = 1$, $|g(z)| < |f_s(z)|$. Also, $f_s(z)$ and $g(z)$ are analytic inside and on $|z| = 1$. Thus by Rouché's theorem, as $f_s(z)$ has one zero in the range $|z| < 1$ (at $z = 0$), so also does $D = f_s(z) + g(z)$ have one zero there.

(c) Since $|\alpha_2(s)| < |\alpha_1(s)|$, the one zero of D of interest must be $\alpha_2(s)$. By the analyticity of $P^*(z, s)$ for $|z| < 1$ ($\text{Re}(s) > 0$), $\alpha_2(s)$ is a zero of the numerator. Hence

$$[\alpha_2(s)]^{i+1} - \mu[1 - \alpha_2(s)]P_0^*(s) = 0$$

Therefore

$$P_0^*(s) = \frac{[\alpha_2(s)]^{i+1}}{\mu[1 - \alpha_2(s)]}$$

(d) We have

$$P^*(z, s) = \frac{z^{i+1} - \mu(1 - z)\dfrac{\alpha_2^{i+1}}{\mu(1 - \alpha_2)}}{-\lambda(z - \alpha_1)(z - \alpha_2)}$$

$$= \frac{(1 - \alpha_2)z^{i+1} - (1 - z)\alpha_2^{i+1}}{-\lambda(z - \alpha_1)(z - \alpha_2)(1 - \alpha_2)}$$

The numerator can be written as

$$z^{i+1} - \alpha_2^{i+1} - \alpha_2 z(z^i - \alpha_2^i)$$

$$= (z - \alpha_2)(z^i + z^{i-1}\alpha_2 + \cdots + z\alpha_2^{i-1} + \alpha_2^i)$$

$$\quad - \alpha_2 z(z - \alpha_2)(z^{i-1} + z^{i-2}\alpha_2 + \cdots + z\alpha_2^{i-2} + \alpha_2^{i-1})$$

$$= (z - \alpha_2)[(1 - \alpha_2)(z^i + z^{i-1}\alpha_2 + \cdots + z\alpha_2^{i-1} + \alpha_2^i) + \alpha_2^{i+1}]$$

Thus

$$P^*(z, s) = \frac{(z - \alpha_2)[(1 - \alpha_2)(z^i + \alpha_2 z^{i-1} + \cdots + \alpha_2^i) + \alpha_2^{i+1}]}{-\lambda(z - \alpha_2)(z - \alpha_1)(1 - \alpha_2)}$$

$$P^*(z, s) = \frac{(z^i + \alpha_2 z^{i-1} + \cdots + \alpha_2^i) + \alpha_2^{i+1}/(1 - \alpha_2)}{\lambda\alpha_1(1 - z/\alpha_1)}$$

(e) We may rewrite

$$P^*(z,s) = \frac{(z^i + \alpha_2 z^{i-1} + \cdots + \alpha_2^i) + \alpha_2^{i+1}/(1 - \alpha_2)}{\lambda \alpha_1} \sum_{k=0}^{\infty} \left(\frac{z}{\alpha_1}\right)^k$$

From this power series, we recognize, for $k \geq i$,

$$P_k^*(s) = \frac{1}{\lambda \alpha_1} \left[\left(\frac{1}{\alpha_1}\right)^{k-i} + \alpha_2 \left(\frac{1}{\alpha_1}\right)^{k-i+1} + \cdots + \alpha_2^i \left(\frac{1}{\alpha_1}\right)^k \right]$$

$$+ \frac{\alpha_2^{i+1}}{1 - \alpha_2} \cdot \frac{1}{\lambda \alpha_1^{k+1}}$$

Now, using the fact that $|\alpha_2| < 1$,

$$P_k^*(s) = \frac{1}{\lambda \alpha_1} \left[\left(\frac{1}{\alpha_1}\right)^{k-i} + \alpha_2 \left(\frac{1}{\alpha_1}\right)^{k-i+1} + \cdots + \alpha_2^i \left(\frac{1}{\alpha_1}\right)^k \right]$$

$$+ \frac{1}{\lambda} \frac{\alpha_2^{i+1}}{\alpha_1^{k+1}} \left[1 + \alpha_2 + \alpha_2^2 + \cdots \right]$$

and thus

$$P_k^*(s) = \frac{1}{\lambda} \left[\left(\frac{1}{\alpha_1}\right)^{k-i+1} + \alpha_1 \alpha_2 \left(\frac{1}{\alpha_1}\right)^{k-i+3} + \cdots + (\alpha_1 \alpha_2)^i \left(\frac{1}{\alpha_1}\right)^{k+i+1} \right]$$

$$+ \frac{1}{\lambda} \frac{\alpha_2^{k+i+2}}{(\alpha_1 \alpha_2)^{k+1}} \left[1 + \alpha_2 + \alpha_2^2 + \cdots \right]$$

Since $\alpha_1 \alpha_2 = \mu/\lambda$, we have $\alpha_2 = \mu/\lambda\alpha_1$. Thus, for $k \geq i$,

$$P_k^*(s) = \frac{1}{\lambda} \left[\alpha_1^{i-k-1} + \left(\frac{\mu}{\lambda}\right) \alpha_1^{i-k-3} + \left(\frac{\mu}{\lambda}\right)^2 \alpha_1^{i-k-5} + \cdots \right.$$

$$\left. + \left(\frac{\mu}{\lambda}\right)^i \alpha_1^{-i-k-1} + \left(\frac{\lambda}{\mu}\right)^{k+1} \sum_{j=k+i+2}^{\infty} \left(\frac{\mu}{\lambda\alpha_1}\right)^j \right]$$

(f) From entry 4 in Table 1.2 and the transform pair given in part (f) of the problem statement, we have

$$\alpha_1^{-k} = \left[\frac{s + \mu + \lambda + \sqrt{(s + \mu + \lambda)^2 - 4\lambda\mu}}{2\lambda} \right]^{-k} \Leftrightarrow e^{-(\lambda+\mu)t} k \rho^{k/2} t^{-1} I_k(at)$$

where $\rho = \lambda/\mu$ and $a = 2\mu\rho^{1/2} = 2\sqrt{\lambda\mu}$. Inverting the expression for $P_k^*(s)$ from part (e), using the fact that

$$\frac{2k}{at} I_k(at) = I_{k-1}(at) - I_{k+1}(at),$$

and carefully reducing the result, will yield Eq. (1.77) for $k \geq i$. Details can be found in [SAAT 61], pages 91–93.

(g) For $k < i$, the value of $P_k^*(s)$ must be found, and the result must be inverted. Further details are on page 93 in [SAAT 61]. $\qquad$ □

PROBLEM 2.21

The random variables $X_1, X_2, \ldots, X_i, \ldots$ are independent and identically distributed each with density $f_X(x)$ and Laplace transform $X^*(s) = E[e^{-sX}]$. Consider a Poisson process $N(t)$ with parameter λ, which is independent of the random variables X_i. Consider now a second random process of the form

$$X(t) = \sum_{i=1}^{N(t)} X_i$$

This second random process is clearly a family of staircase functions, where the jumps occur at the discontinuities of the random process $N(t)$; the magnitudes of such jumps are given by the random variables X_i. Show that the Laplace transform of this second random process, namely, $X^*(t, s) = E[e^{-sX(t)}]$, is given by

$$X^*(t, s) = e^{\lambda t[X^*(s) - 1]}$$

SOLUTION

Condition on $N(t) = n$. Then the (conditional) transform is given as follows:

$$X^*(t, s \mid N(t) = n) = E[e^{-sX(t)} \mid N(t) = n] = E\left[e^{-s\sum_{i=1}^{n} X_i}\right]$$

$$= (E[e^{-sX_1}])^n \qquad \text{(as the } X_i \text{ are i.i.d.)}$$

$$= [X^*(s)]^n$$

Since $N(t)$ is Poisson distributed with parameter λt, we may uncondition as follows (note that we assume $P[N(0) = 0] = 1$):

$$X^*(t, s) = \sum_{n=0}^{\infty} X^*(t, s \mid N(t) = n) \cdot P[N(t) = n]$$

$$= \sum_{n=0}^{\infty} [X^*(s)]^n \frac{(\lambda t)^n e^{-\lambda t}}{n!} = e^{-\lambda t} e^{\lambda t X^*(s)}$$

$$= e^{\lambda t[X^*(s) - 1]} \qquad \qquad \square$$

PROBLEM 2.22

Passengers and taxis arrive at a service point from independent Poisson processes at rates λ, μ, respectively. Let the queue size at time t be q_t, a negative value denoting a line of taxis, a positive value denoting a queue of passengers. Show that, starting with $q_0 = 0$, the distribution of q_t is given by the difference between independent Poisson variables of means λt, μt. Show by using the normal approximation (see [KLEI 75]) that if $\lambda = \mu$, the probability that $-k \leq q_t \leq k$ is, for large t, $(2k + 1)(4\pi\lambda t)^{-1/2}$.

SOLUTION

Let $P_n(t) = P[q_t = n]$, $-\infty < n < \infty$. For $n < 0$, we have n taxis waiting. For $n > 0$, we have n passengers waiting. For $n = 0$, we have no queue. The equations of motion are simply

$$\frac{dP_n(t)}{dt} = -(\lambda + \mu)P_n(t) + \mu P_{n+1}(t) + \lambda P_{n-1}(t) \qquad \text{for } -\infty < n < \infty$$

Define $P(z,t) = \sum_{-\infty}^{+\infty} P_n(t)z^n$. Multiplying the nth equation by z^n and summing yields

$$\sum_{-\infty}^{+\infty} \frac{dP_n(t)}{dt} z^n = -(\lambda + \mu) \sum_{-\infty}^{+\infty} P_n(t)z^n + \mu \sum_{-\infty}^{+\infty} P_{n+1}(t)z^n + \lambda \sum_{-\infty}^{+\infty} P_{n-1}(t)z^n$$

$$\frac{\partial P(z,t)}{\partial t} = -P(z,t)\left[\lambda(1 - z) + \mu\left(1 - \frac{1}{z}\right)\right]$$

Solving this partial differential equation gives

$$P(z,t) = A(z)e^{-\lambda t(1-z) - \mu t(1 - \frac{1}{z})}$$

The initial condition $P_0(0) = 1$ implies $P(z,0) = 1$. Hence $A(z) = 1$. Therefore

$$P(z,t) = e^{-\lambda t(1-z) - \mu t(1 - \frac{1}{z})}$$

But the arrival process for passengers has, for fixed t, the z-transform

$$P_1(z,t) = \sum_{n=0}^{\infty} \frac{(\lambda t)^n}{n!} e^{-\lambda t} z^n = e^{-\lambda t(1-z)}$$

and the arrival process for taxis has the corresponding transform

$$P_2(z,t) = \sum_{n=0}^{\infty} \frac{(\mu t)^n}{n!} e^{-\mu t} z^n = e^{-\mu t(1-z)}$$

From these results we see that

$$P(z,t) = P_1(z,t) P_2(1/z,t)$$

Now note for independent random variables X, Y (whose z-transforms are F and G, respectively) and with $W = X - Y$ that

$$H(z) = E[z^W] = E[z^{X-Y}] = E[z^X] \cdot E[z^{-Y}]$$
$$= E[z^X] \cdot E\left[(1/z)^Y\right] = F(z)\,G(1/z)$$

Thus $P_1(z, t) \cdot P_2(1/z, t)$ represents the z-transform of the difference between independent Poisson variables of means $\lambda t, \mu t$. But since it is also equal to the z-transform $P(z, t)$ of q_t, we see that the distribution of q_t has the required form. We recognize that the mean of q_t is $\lambda t - \mu t$, and the variance is $\lambda t + \mu t$. Now assume $\lambda = \mu$. Hence the mean of q_t is 0 with variance $2\lambda t$. Using the normal approximation, we have

$$P[-k \le q_t \le k] \cong \int_{-(k+\frac{1}{2})}^{k+\frac{1}{2}} \frac{1}{\sqrt{2\pi}\,\sqrt{2\lambda t}}\, e^{-\frac{x^2}{4\lambda t}}\, dx$$

But for large t, $e^{-\frac{x^2}{4\lambda t}} \to 1$. So

$$P[-k \le q_t \le k] \cong \int_{-(k+\frac{1}{2})}^{k+\frac{1}{2}} \frac{1}{\sqrt{4\pi\lambda t}}\, dx = \frac{2k+1}{\sqrt{4\pi\lambda t}} \qquad \square$$

CHAPTER 3

BIRTH–DEATH QUEUEING SYSTEMS

PROBLEM 3.1

Consider a birth–death queueing system in which

$$\lambda_k = \begin{cases} \lambda & 0 \le k \le K \\ 2\lambda & K < k \end{cases}$$

$$\mu_k = \mu \qquad k = 1, 2, \ldots$$

(a) Find the equilibrium probabilities p_k for the number in the system.

(b) What relationship must exist among the parameters of the problem in order that a solution exist? Interpret this answer in terms of the possible dynamics of the system.

SOLUTION

(a) **Case (1)**: $0 \le k \le K + 1$

Equation (1.63) gives

$$p_k = p_0 \prod_{i=0}^{k-1} \frac{\lambda}{\mu} = p_0 \left(\frac{\lambda}{\mu} \right)^k$$

Case (2): $k > K + 1$

Equation (1.63) gives

$$p_k = p_0 \prod_{i=0}^{K} \frac{\lambda}{\mu} \prod_{i=K+1}^{k-1} \frac{2\lambda}{\mu}$$

$$p_k = p_0 \left[\left(\frac{\lambda}{\mu}\right)^{K+1} \left(\frac{2\lambda}{\mu}\right)^{k-K-1} \right] = p_0 \frac{1}{2^{K+1}} \left(\frac{2\lambda}{\mu}\right)^{k}$$

Using the conservation of probability relation, namely, $\sum_{k=0}^{\infty} p_k = 1$, we solve for p_0 as follows:

$$1 = p_0 \left[\sum_{k=0}^{K+1} \left(\frac{\lambda}{\mu}\right)^{k} + \sum_{k=K+2}^{\infty} \frac{1}{2^{K+1}} \left(\frac{2\lambda}{\mu}\right)^{k} \right]$$

$$= p_0 \left[\frac{1 - \left(\frac{\lambda}{\mu}\right)^{K+2}}{1 - \frac{\lambda}{\mu}} + \frac{1}{2^{K+1}} \left(\frac{2\lambda}{\mu}\right)^{K+2} \frac{1}{1 - \frac{2\lambda}{\mu}} \right]$$

Thus

$$p_0 = \frac{\left(1 - \frac{\lambda}{\mu}\right)\left(1 - \frac{2\lambda}{\mu}\right)}{1 - \frac{2\lambda}{\mu} + \left(\frac{\lambda}{\mu}\right)^{K+2}}$$

and

$$p_k = \begin{cases} \dfrac{\left(1 - \frac{\lambda}{\mu}\right)\left(1 - \frac{2\lambda}{\mu}\right)}{1 - \frac{2\lambda}{\mu} + \left(\frac{\lambda}{\mu}\right)^{K+2}} \left(\frac{\lambda}{\mu}\right)^{k} & 0 \le k \le K + 1 \\[4ex] \dfrac{\left(1 - \frac{\lambda}{\mu}\right)\left(1 - \frac{2\lambda}{\mu}\right)}{1 - \frac{2\lambda}{\mu} + \left(\frac{\lambda}{\mu}\right)^{K+2}} \left(\frac{2\lambda}{\mu}\right)^{k} \frac{1}{2^{K+1}} & k > K + 1 \end{cases}$$

(b) We must have $(2\lambda/\mu) < 1$ or $2\lambda < \mu$ to ensure that p_k does not produce a divergent series. We observe that if the system goes unstable, then the number in system will exceed K with probability one. Thus the relevant birth parameter is $\lambda_k = 2\lambda$, and our stability condition is $2\lambda < \mu$. □

PROBLEM 3.2

Consider a birth–death queueing system in which

$$\lambda_k = \alpha^k \lambda \qquad k \geq 0, 0 \leq \alpha < 1$$
$$\mu_k = \mu \qquad k \geq 1$$

(a) Find the equilibrium probability p_k of having k customers in the system. Express your answer in terms of p_0.

(b) Give an expression for p_0.

SOLUTION

(a) From Eq. (1.63) we have

$$p_k = p_0 \prod_{i=0}^{k-1} \alpha^i \left(\frac{\lambda}{\mu} \right) = p_0 \left(\frac{\lambda}{\mu} \right)^k \alpha^{\sum_{i=0}^{k-1} i}$$

$$p_k = p_0 \left(\frac{\lambda}{\mu} \right)^k \alpha^{\frac{(k-1)k}{2}}$$

Therefore

$$p_k = p_0 \left[\frac{\lambda \alpha^{(k-1)/2}}{\mu} \right]^k$$

(b) Using conservation of probability, we find

$$\sum_{k=0}^{\infty} p_k = 1 = p_0 \sum_{k=0}^{\infty} \left[\frac{\lambda \alpha^{(k-1)/2}}{\mu} \right]^k$$

So

$$p_0 = \frac{1}{\sum_{k=0}^{\infty} \left[\frac{\lambda \alpha^{(k-1)/2}}{\mu} \right]^k}$$

Note for $0 \leq \alpha < 1$, this system is *always* stable. □

PROBLEM 3.3

Consider an M/M/2 queueing system where the average arrival rate is λ customers per second and the average service time is $1/\mu$ sec, where $\lambda < 2\mu$.

(a) Find the differential equations that govern the time-dependent probabilities $P_k(t)$.

(b) Find the equilibrium probabilities

$$p_k = \lim_{t \to \infty} P_k(t)$$

SOLUTION

The birth and death coefficients are

$$\lambda_k = \lambda \quad k \geq 0, \qquad \mu_k = \begin{cases} \mu & k = 1 \\ 2\mu & k \geq 2 \end{cases}$$

(a) The differential-difference equations are

$$\frac{dP_0(t)}{dt} = -\lambda P_0(t) + \mu P_1(t) \qquad k = 0$$

$$\frac{dP_1(t)}{dt} = -(\lambda + \mu)P_1(t) + \lambda P_0(t) + 2\mu P_2(t) \qquad k = 1$$

$$\frac{dP_k(t)}{dt} = -(\lambda + 2\mu)P_k(t) + \lambda P_{k-1}(t) + 2\mu P_{k+1}(t) \qquad k \geq 2$$

(b) Using Eq. (1.63) we have

$$p_k = p_0 \left(\frac{\lambda}{\mu}\right)\left(\frac{\lambda}{2\mu}\right)^{k-1} = 2p_0\left(\frac{\lambda}{2\mu}\right)^k \qquad \text{for } k \geq 1$$

Using $\sum_{k=0}^{\infty} p_k = 1$ we have

$$p_0 + 2p_0 \sum_{k=1}^{\infty}\left(\frac{\lambda}{2\mu}\right)^k = 1$$

$$p_0\left[1 + \frac{\lambda}{\mu}\frac{1}{1 - \dfrac{\lambda}{2\mu}}\right] = 1$$

$$p_0 = \frac{1 - \dfrac{\lambda}{2\mu}}{1 + \dfrac{\lambda}{2\mu}}$$

$$p_k = \frac{1 - \dfrac{\lambda}{2\mu}}{1 + \dfrac{\lambda}{2\mu}} 2\left(\frac{\lambda}{2\mu}\right)^k \qquad k \geq 1 \qquad \square$$

PROBLEM 3.4

Consider an M/M/1 system with parameters λ, μ in which customers are impatient. Specifically, upon arrival, customers estimate their queueing time w and then join the queue with probability $e^{-\alpha w}$ (or leave with probability $1 - e^{-\alpha w}$). The estimate is $w = k/\mu$ when the new arrival finds k in the system. Assume $0 \leq \alpha$.

(a) In terms of p_0, find the equilibrium probabilities p_k of finding k in the system. Give an expression for p_0 in terms of the system parameters.

(b) Under what conditions will the equilibrium solution hold (i.e., when will $p_0 > 0$)?

(c) For $\alpha \to \infty$, find p_k explicitly and find the average number in the system.

SOLUTION

The birth and death coefficients are

$$\lambda_k = \lambda e^{-\frac{\alpha k}{\mu}}, \ \mu_k = \mu$$

(a) Equation (1.63) gives

$$p_k = p_0 \prod_{i=0}^{k-1} \frac{\lambda e^{-\frac{\alpha i}{\mu}}}{\mu} = p_0 \left(\frac{\lambda}{\mu} \right)^k e^{-\frac{\alpha}{\mu} \sum_{i=0}^{k-1} i}$$

$$p_k = p_0 \left(\frac{\lambda}{\mu} \right)^k e^{-\frac{\alpha k(k-1)}{2\mu}}$$

$$\sum_{k=0}^{\infty} p_k = 1 = p_0 \sum_{k=0}^{\infty} \left(\frac{\lambda}{\mu} \right)^k e^{-\frac{\alpha k(k-1)}{2\mu}}$$

$$p_0 = \frac{1}{\sum_{k=0}^{\infty} \left(\frac{\lambda}{\mu} \right)^k e^{-\frac{\alpha k(k-1)}{2\mu}}}$$

(b) The denominator for p_0 converges if either $0 < \alpha$ and $0 < \mu$ or if $\alpha = 0$ and $\lambda < \mu$.

(c) For $\alpha \to \infty$, $p_k \to 0$ for $k \geq 2$. Thus we only move between the two states 0 and 1 with $\lambda_0 = \lambda$, $\lambda_1 = 0$ and $\mu_1 = \mu$. Thus solving for p_0 and p_1 (see also Problem 2.11 for $P_k(t)$) gives

$$p_0 = \frac{\mu}{\lambda + \mu}$$

$$p_1 = \frac{\lambda}{\lambda + \mu}$$

$$\overline{N} = 0 \cdot p_0 + 1 \cdot p_1 = \frac{\lambda}{\lambda + \mu} \qquad \qquad \square$$

PROBLEM 3.5

Consider a birth–death system with the following birth and death coefficients:

$$\lambda_k = (k + 2)\lambda \qquad k = 0, 1, 2, \ldots$$

$$\mu_k = k\mu \qquad k = 1, 2, 3, \ldots$$

All other coefficients are zero.

(a) Solve for p_k. Be sure to express your answer explicitly in terms of λ, k, and μ only.

(b) Find the average number of customers in the system.

SOLUTION

(a) Equation (1.63) gives

$$p_k = p_0 \left(\frac{\lambda}{\mu}\right)^k \frac{2 \cdot 3 \cdots (k+1)}{1 \cdot 2 \cdots k}$$

$$p_k = p_0 (k+1) \left(\frac{\lambda}{\mu}\right)^k \qquad \text{for } k \geq 0$$

$$1 = \sum_{k=0}^{\infty} p_k = p_0 \sum_{k=0}^{\infty} (k+1) \left(\frac{\lambda}{\mu}\right)^k$$

Here we demonstrate the "differentiation trick" for summing series. Since

$$\sum_{k=0}^{\infty} (k+1) \left(\frac{\lambda}{\mu}\right)^k = \frac{\partial}{\partial \left(\frac{\lambda}{\mu}\right)} \sum_{k=0}^{\infty} \left(\frac{\lambda}{\mu}\right)^{k+1}$$

we have

$$\sum_{k=0}^{\infty} (k+1) \left(\frac{\lambda}{\mu}\right)^k = \frac{\partial}{\partial \left(\frac{\lambda}{\mu}\right)} \left[\frac{\frac{\lambda}{\mu}}{1 - \frac{\lambda}{\mu}}\right] = \frac{1}{\left(1 - \frac{\lambda}{\mu}\right)^2}$$

Thus

$$p_0 = \left(1 - \frac{\lambda}{\mu}\right)^2$$

and so

$$p_k = \left(1 - \frac{\lambda}{\mu}\right)^2 (k+1) \left(\frac{\lambda}{\mu}\right)^k \qquad k \geq 0$$

(b) We have

$$\bar{N} = \sum_{k=0}^{\infty} k p_k = \sum_{k=1}^{\infty} k \left(1 - \frac{\lambda}{\mu}\right)^2 (k+1) \left(\frac{\lambda}{\mu}\right)^k$$

$$= \left(1 - \frac{\lambda}{\mu}\right)^2 \frac{\lambda}{\mu} \frac{\partial^2}{\partial \left(\frac{\lambda}{\mu}\right)^2} \left[\sum_{k=0}^{\infty} \left(\frac{\lambda}{\mu}\right)^{k+1}\right]$$

$$\overline{N} = \left(1 - \frac{\lambda}{\mu}\right)^2 \frac{\lambda}{\mu} \frac{\partial}{\partial \left(\frac{\lambda}{\mu}\right)} \left[\frac{1}{\left(1 - \frac{\lambda}{\mu}\right)^2}\right] = \left(1 - \frac{\lambda}{\mu}\right)^2 \frac{\lambda}{\mu} \frac{2}{\left(1 - \frac{\lambda}{\mu}\right)^3}$$

$$\overline{N} = \frac{2\left(\frac{\lambda}{\mu}\right)}{1 - \frac{\lambda}{\mu}} \qquad \qquad \square$$

PROBLEM 3.6

Consider a birth–death process with the following coefficients:

$$\lambda_k = \alpha k(K_2 - k) \qquad k = K_1, K_1 + 1, \ldots, K_2$$
$$\mu_k = \beta k(k - K_1) \qquad k = K_1, K_1 + 1, \ldots, K_2$$

where $K_1 \le K_2$ and where these coefficients are zero outside the range $K_1 \le k \le K_2$. Solve for p_k (assuming that the system initially contains $K_1 \le k \le K_2$ customers).

SOLUTION

Clearly $p_k = 0$ for $k < K_1$, $k > K_2$. Using the obvious translation of Eq. (1.63) in the range $K_1 \le k \le K_2$, we get

$$p_k = p_{K_1} \prod_{i=K_1}^{k-1} \frac{\alpha i(K_2 - i)}{\beta(i + 1)(i + 1 - K_1)}$$

$$p_k = p_{K_1} \left(\frac{\alpha}{\beta}\right)^{k-K_1} \frac{K_1(K_1 + 1) \cdots (k - 1)}{(K_1 + 1)(K_1 + 2) \cdots k} \cdot \frac{(K_2 - K_1) \cdots (K_2 - k + 1)}{1 \cdot 2 \cdots (k - K_1)}$$

Multiplying the top and bottom of the right-hand expression by $K_1! \, (K_2 - k)!$ we find

$$p_k = p_{K_1} \left(\frac{\alpha}{\beta}\right)^{k-K_1} \frac{(k - 1)! \, K_1(K_2 - K_1)!}{(k - 1)! \, k(K_2 - k)! \, (k - K_1)!}$$

$$p_k = p_{K_1} \left(\frac{\alpha}{\beta}\right)^{k-K_1} \frac{K_1}{k} \binom{K_2 - K_1}{k - K_1} \qquad k = K_1, \ldots, K_2$$

where, by conserving probability, we get

$$p_{K_1} = \frac{1}{\displaystyle\sum_{k=K_1}^{K_2} \left(\frac{\alpha}{\beta}\right)^{k-K_1} \frac{K_1}{k} \binom{K_2 - K_1}{k - K_1}} \qquad \qquad \square$$

PROBLEM 3.7

Consider an M/M/m system that is to serve the pooled sum of two Poisson arrival streams; the ith stream has an average arrival rate given by λ_i and exponentially distributed service times with mean $1/\mu_i$ ($i = 1, 2$). The first stream is an ordinary stream whereby each arrival requires exactly one of the m servers; if all m servers are busy then any newly arriving customer of type 1 is lost. Customers from the second class each require the simultaneous use of m_0 servers (and will occupy them all simultaneously for the same exponentially distributed amount of time whose mean is $1/\mu_2$ sec); if a customer from this class finds less than m_0 idle servers then he too is lost to the system. Find the fraction of type 1 customers and the fraction of type 2 customers that are lost.

SOLUTION

The state space consists of (a finite number of) two-dimensional states (i, j), where (i, j) represents the system state in which i customers of type 1 and j customers of type 2 are present. Let n and k be such that $m = nm_0 + k, 0 \le k \le m_0 - 1$. The set of states consists of all (i, j) such that $0 \le i \le m, 0 \le j \le n$, and $i + jm_0 \le m$ (each customer of type 2 requires m_0 servers). The state diagram is drawn below:

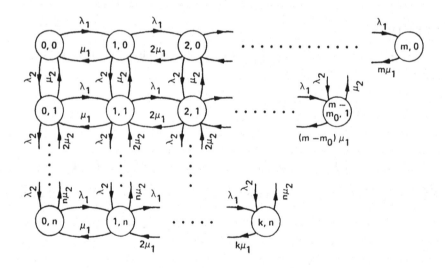

The general balance equation (Rate In = Rate Out) for an "internal" state is

$$\lambda_2 p(i, j - 1) + \lambda_1 p(i - 1, j) + (j + 1)\mu_2 p(i, j + 1) + (i + 1)\mu_1 p(i + 1, j)$$
$$= j\mu_2 p(i, j) + i\mu_1 p(i, j) + \lambda_2 p(i, j) + \lambda_1 p(i, j)$$

This set of equations, supplemented with the boundary equations and the conservation of probability, has a unique solution (since we have an ergodic Markov chain). Thus, if we can guess the $p(i, j)$ and show that they satisfy those equations, then we know we have the equilibrium solution. The reader should show that each row and each column of our state diagram is the same as that of an M/M/m/m loss system. Therefore, we choose to guess the two-dimensional "product form" solution as suggested by Eq. (1.88) for M/M/m/m:

$$p(i, j) = \frac{1}{i!} \left(\frac{\lambda_1}{\mu_1} \right)^i \frac{1}{j!} \left(\frac{\lambda_2}{\mu_2} \right)^j p(0, 0)$$

Checking this proposed solution for an internal state, we have

$$
\frac{\text{Rate In}}{p(0,0)} = \lambda_2 \frac{1}{i!} \left(\frac{\lambda_1}{\mu_1} \right)^i \frac{1}{(j-1)!} \left(\frac{\lambda_2}{\mu_2} \right)^{j-1} + \lambda_1 \frac{1}{(i-1)!} \left(\frac{\lambda_1}{\mu_1} \right)^{i-1} \frac{1}{j!} \left(\frac{\lambda_2}{\mu_2} \right)^j
$$

$$
+ (j+1)\mu_2 \frac{1}{i!} \left(\frac{\lambda_1}{\mu_1} \right)^i \frac{1}{(j+1)!} \left(\frac{\lambda_2}{\mu_2} \right)^{j+1}
$$

$$
+ (i+1)\mu_1 \frac{1}{(i+1)!} \left(\frac{\lambda_1}{\mu_1} \right)^{i+1} \frac{1}{j!} \left(\frac{\lambda_2}{\mu_2} \right)^j
$$

$$
= (j\mu_2 + i\mu_1 + \lambda_2 + \lambda_1) \frac{1}{i!} \left(\frac{\lambda_1}{\mu_1} \right)^i \frac{1}{j!} \left(\frac{\lambda_2}{\mu_2} \right)^j
$$

$$
= \frac{\text{Rate Out}}{p(0,0)}
$$

which confirms our proposed solution. (Note: The boundary states should also be checked.) Using conservation of probability to evaluate $p(0, 0)$ we have the complete solution:

$$
p(i, j) = \frac{\dfrac{1}{i!} \left(\dfrac{\lambda_1}{\mu_1} \right)^i \dfrac{1}{j!} \left(\dfrac{\lambda_2}{\mu_2} \right)^j}{\displaystyle\sum_{j=0}^{n} \frac{1}{j!} \left(\frac{\lambda_2}{\mu_2} \right)^j \sum_{i=0}^{m-jm_0} \frac{1}{i!} \left(\frac{\lambda_1}{\mu_1} \right)^i}
$$

Let us now find the loss probabilities. A customer of type 1 is lost iff all m servers are busy. This occurs if the system is in the rightmost state of any row. Summing the probabilities of these $n + 1$ states, we have

$$
P[\text{type 1 lost}] = p(0, 0) \left[\frac{1}{m!} \left(\frac{\lambda_1}{\mu_1} \right)^m + \frac{1}{(m - m_0)!} \left(\frac{\lambda_1}{\mu_1} \right)^{m - m_0} \left(\frac{\lambda_2}{\mu_2} \right) + \cdots \right.
$$

$$
\left. + \frac{1}{(m - nm_0)!} \left(\frac{\lambda_1}{\mu_1} \right)^{m - nm_0} \frac{1}{n!} \left(\frac{\lambda_2}{\mu_2} \right)^n \right]
$$

$$P[\text{type 1 lost}] = \frac{\sum_{j=0}^{n} \frac{1}{j!} \left(\frac{\lambda_2}{\mu_2}\right)^j \frac{1}{(m-jm_0)!} \left(\frac{\lambda_1}{\mu_1}\right)^{m-jm_0}}{\sum_{j=0}^{n} \frac{1}{j!} \left(\frac{\lambda_2}{\mu_2}\right)^j \sum_{i=0}^{m-jm_0} \frac{1}{i!} \left(\frac{\lambda_1}{\mu_1}\right)^i}$$

A type 2 customer is lost iff less than m_0 servers are idle; that is, at least $m - m_0 + 1$ servers are busy.

$$P[\text{type 2 lost}] = \sum_{i=m-m_0+1}^{m} p(i,0) + \sum_{i=m-2m_0+1}^{m-m_0} p(i,1) + \cdots$$

$$+ \sum_{i=m-nm_0+1}^{m-(n-1)m_0} p(i,n-1) + \sum_{i=0}^{k} p(i,n)$$

Summing these $m + 1$ probabilities we get

$$P[\text{type 2 lost}] = \frac{\sum_{j=0}^{n-1} \frac{1}{j!} \left(\frac{\lambda_2}{\mu_2}\right)^j \sum_{i=m-(j+1)m_0+1}^{m-jm_0} \frac{1}{i!} \left(\frac{\lambda_1}{\mu_1}\right)^i}{\sum_{j=0}^{n} \frac{1}{j!} \left(\frac{\lambda_2}{\mu_2}\right)^j \sum_{i=0}^{m-jm_0} \frac{1}{i!} \left(\frac{\lambda_1}{\mu_1}\right)^i}$$

$$+ \frac{\frac{1}{n!} \left(\frac{\lambda_2}{\mu_2}\right)^n \sum_{i=0}^{k} \frac{1}{i!} \left(\frac{\lambda_1}{\mu_1}\right)^i}{\sum_{j=0}^{n} \frac{1}{j!} \left(\frac{\lambda_2}{\mu_2}\right)^j \sum_{i=0}^{m-jm_0} \frac{1}{i!} \left(\frac{\lambda_1}{\mu_1}\right)^i} \qquad \square$$

PROBLEM 3.8

Consider a finite customer population system with a single server (i.e., the system M/M/1/∞/M). Let the parameters M, λ be replaced by M, λ'. It can be shown that if $M \to \infty$ and $\lambda' \to 0$ such that $\lim M\lambda' = \lambda$, then the finite population system becomes an infinite population system with exponential interarrival times (at a mean rate of λ customers per second). Now consider the case of an M/M/m system with a finite population of size M and a finite storage of size K, that is, the system M/M/m/K/M. The parameters of this system are now to be denoted M, λ', m, μ, K in the obvious way. Show what value these parameters must take on if they are to represent each of the following cases:

(i) M/M/1

(ii) M/M/∞

(iii) M/M/m
(iv) M/M/1/K
(v) M/M/m/m
(vi) M/M/1/∞/M
(vii) M/M/∞/∞/M

SOLUTION

The solution to our M/M/m/K/M system, which is a birth–death system, may be obtained by applying the general solution given in Eq. (1.63). We assume that $M \geq K \geq m$. All that is necessary is to find expressions for λ_k and μ_k. For λ_k, we see that, when k customers are in the system (queue plus service), there are $M - k$ customers, each arriving at a rate λ' and therefore a total arrival rate $\lambda_k = \lambda'(M - k)$. However, customers arriving to find K already in the system are "lost" and return immediately to the arriving state as if they had just completed service. This leads to the following set of birth coefficients:

$$\lambda_k = \begin{cases} \lambda'(M - k) & 0 \leq k \leq K - 1 \\ 0 & \text{otherwise} \end{cases}$$

For μ_k, we see that, if $k \leq m$, the death rate will be $k\mu$; however, if $k > m$, only m customers will be in service. This leads to the following set of death coefficients:

$$\mu_k = \begin{cases} k\mu & 0 \leq k \leq m \\ m\mu & k \geq m \end{cases}$$

Applying these coefficients to Eq. (1.63), we see the solution breaks into two regions, namely, for the range $0 \leq k \leq m - 1$ we have

$$p_k = p_0 \prod_{i=0}^{k-1} \frac{\lambda'(M - i)}{(i + 1)\mu}$$

$$= p_0 \left(\frac{\lambda'}{\mu}\right)^k \binom{M}{k} \qquad 0 \leq k \leq m - 1$$

For the region $m \leq k \leq K$ we have

$$p_k = p_0 \prod_{i=0}^{m-1} \frac{\lambda'(M - i)}{(i + 1)\mu} \prod_{i=m}^{k-1} \frac{\lambda'(M - i)}{m\mu}$$

$$= p_0 \left(\frac{\lambda'}{\mu}\right)^k \binom{M}{k} \frac{k!}{m!} m^{m-k} \qquad m \leq k \leq K$$

Here p_0 is found from the conservation of probability, namely, $\sum_{k=0}^{K} p_k = 1$. We must specialize the parameters and take the limit $M\lambda' = \lambda$ as $M \to \infty$, $\lambda' \to 0$ of these equations for each case below:

(i) M/M/1

$$m = 1, \ K = \infty, \ M = \infty, \ \lambda' = \lambda/M$$

For finite M, our solution gives

$$p_k = p_0 \left(\frac{\lambda}{M\mu}\right)^k \binom{M}{k} k! = p_0 \left(\frac{\lambda}{\mu}\right)^k \frac{M \cdot (M - 1) \cdots (M - k + 1)}{M^k}$$

Letting $M \to \infty$,

$$\lim_{M \to \infty} p_k = p_0 \left(\frac{\lambda}{\mu}\right)^k \qquad \text{[as in Eq. (1.65)]}$$

(ii) M/M/∞

$$m = \infty, \ K = \infty, \ M = \infty, \ \lambda' = \lambda/M$$

For finite M, our solution gives

$$p_k = p_0 \left(\frac{\lambda}{M\mu}\right)^k \binom{M}{k} = p_0 \left(\frac{\lambda}{\mu}\right)^k \frac{M \cdot (M - 1) \cdots (M - k + 1)}{M^k k!}$$

Letting $M \to \infty$,

$$\lim_{M \to \infty} p_k = p_0 \left(\frac{\lambda}{\mu}\right)^k \frac{1}{k!} \qquad \text{[as in Eq. (1.89)]}$$

(iii) M/M/m

$$m = m, \ K = \infty, \ M = \infty, \ \lambda' = \lambda/M$$

For $k \leq m - 1$ and finite M, our solution gives

$$p_k = p_0 \left(\frac{\lambda}{M\mu}\right)^k \binom{M}{k} = p_0 \left(\frac{\lambda}{\mu}\right)^k \frac{M \cdot (M - 1) \cdots (M - k + 1)}{M^k k!}$$

Letting $M \to \infty$,

$$\lim_{M \to \infty} p_k = p_0 \left(\frac{\lambda}{\mu}\right)^k \frac{1}{k!} = p_0 \frac{(m\rho)^k}{k!} \qquad \text{where } \rho = \frac{\lambda}{m\mu} \quad \text{[as in Eq. (1.86)]}$$

For $k \geq m$ and finite M, our solution gives

$$p_k = p_0 \left(\frac{\lambda}{M\mu}\right)^k \frac{M \cdot (M - 1) \cdots (M - k + 1)k!}{k! \, m!} m^{m-k}$$

Letting $M \to \infty$,

$$\lim_{M \to \infty} p_k = p_0 \left(\frac{\lambda}{\mu}\right)^k \frac{m^{m-k}}{m!} = p_0 \rho^k \frac{m^m}{m!} \qquad \text{[as in Eq. (1.86)]}$$

(iv) M/M/1/K

$$m = 1, \ K = K, \ M = \infty, \ \lambda' = \lambda/M$$

For finite M, our solution gives

$$p_k = p_0 \left(\frac{\lambda}{M\mu} \right)^k \frac{M \cdot (M-1) \cdots (M-k+1)}{k!} k! \qquad 1 \le k \le K$$

Letting $M \to \infty$,

$$\lim_{M \to \infty} p_k = p_0 \left(\frac{\lambda}{\mu} \right)^k \qquad \text{for } k \le K$$

(v) M/M/m/m

$$m = m, \ K = m, \ M = \infty, \ \lambda' = \lambda/M$$

For finite M, our solution gives

$$p_k = p_0 \left(\frac{\lambda}{M\mu} \right)^k \frac{M \cdot (M-1) \cdots (M-k+1)}{k!} \qquad 0 \le k \le m$$

Letting $M \to \infty$,

$$\lim_{M \to \infty} p_k = p_0 \left(\frac{\lambda}{\mu} \right)^k \frac{1}{k!} \qquad 0 \le k \le m \quad \text{[as in Eq. (1.88)]}$$

(vi) M/M/1/∞/M

$$m = 1, \ K = M, \ M = M, \ \lambda' = \lambda$$

Our solution gives

$$p_k = p_0 \left(\frac{\lambda}{\mu} \right)^k \frac{M!}{(M-k)! \, k!} k!$$

$$p_k = p_0 \left(\frac{\lambda}{\mu} \right)^k \frac{M!}{(M-k)!} \qquad 0 \le k \le M \quad \text{[as in Eq. (1.85)]}$$

(vii) M/M/∞/∞/M

$$m = M, \ K = M, \ M = M, \ \lambda' = \lambda$$

Our solution gives

$$p_k = p_0 \left(\frac{\lambda}{\mu} \right)^k \binom{M}{k} \qquad 0 \le k \le M \qquad \qquad \square$$

PROBLEM 3.9

Using the definition for $B(m, \lambda/\mu)$ and $C(m, \lambda/\mu)$ in Section 1.5, establish the following for $\lambda/\mu > 0$, $m = 1, 2, \ldots$:

(a) $B\left(m, \dfrac{\lambda}{\mu}\right) < \displaystyle\sum_{k=m}^{\infty} \dfrac{(\lambda/\mu)^k}{k!} e^{-\lambda/\mu} < C\left(m, \dfrac{\lambda}{\mu}\right)$

(b) $C\left(m, \dfrac{\lambda}{\mu}\right) = \dfrac{B\left(m, \dfrac{\lambda}{\mu}\right)}{1 - \dfrac{\lambda}{m\mu}\left[1 - B\left(m, \dfrac{\lambda}{\mu}\right)\right]}$

(c) $B\left(m + 1, \dfrac{\lambda}{\mu}\right) = \dfrac{\dfrac{\lambda}{\mu} B\left(m, \dfrac{\lambda}{\mu}\right)}{m + 1 + \dfrac{\lambda}{\mu} B\left(m, \dfrac{\lambda}{\mu}\right)}$

SOLUTION

(a) We begin with the obvious inequality:

$$\frac{(\lambda/\mu)^m}{m!} < \sum_{k=m}^{\infty} \frac{(\lambda/\mu)^k}{k!}$$

$$\frac{(\lambda/\mu)^m}{m!}\left[\sum_{k=0}^{m-1} \frac{(\lambda/\mu)^k}{k!}\right] < \left[\sum_{k=m}^{\infty} \frac{(\lambda/\mu)^k}{k!}\right]\left[\sum_{k=0}^{m-1} \frac{(\lambda/\mu)^k}{k!}\right]$$

$$\frac{(\lambda/\mu)^m}{m!}\left[e^{\lambda/\mu} - \sum_{k=m}^{\infty} \frac{(\lambda/\mu)^k}{k!}\right] < \left[\sum_{k=m}^{\infty} \frac{(\lambda/\mu)^k}{k!}\right]\left[\sum_{k=0}^{m-1} \frac{(\lambda/\mu)^k}{k!}\right]$$

$$e^{\lambda/\mu}\frac{(\lambda/\mu)^m}{m!} < \left[\sum_{k=m}^{\infty} \frac{(\lambda/\mu)^k}{k!}\right]\left[\sum_{k=0}^{m} \frac{(\lambda/\mu)^k}{k!}\right]$$

$$\frac{(\lambda/\mu)^m/m!}{\sum_{k=0}^{m}(\lambda/\mu)^k/k!} = B\left(m, \frac{\lambda}{\mu}\right) < \sum_{k=m}^{\infty} \frac{(\lambda/\mu)^k}{k!} e^{-\lambda/\mu}$$

This establishes the left-hand inequality of (a). For the right-hand inequality, we proceed as follows. For $k > m$,

$$k! = m!\,(m + 1)\cdots(k) > m!\,m^{k-m}$$

So

$$\frac{1}{k!} < \frac{1}{m!}\frac{1}{m^{k-m}}$$

$$\sum_{k=m}^{\infty} \frac{(\lambda/\mu)^k}{k!} < \sum_{k=m}^{\infty} \frac{(\lambda/\mu)^k}{m!}\frac{1}{m^{k-m}}$$

Letting $\rho = \lambda/m\mu$, and shifting the index of summation, we have (for $\rho < 1$)

$$\sum_{k=m}^{\infty} \frac{(\lambda/\mu)^k}{k!} < \frac{(\lambda/\mu)^m}{m!}\sum_{j=0}^{\infty}\rho^j = \left(\frac{(m\rho)^m}{m!}\right)\left(\frac{1}{1-\rho}\right)$$

We now proceed as for the left-hand inequality.

$$\left[\sum_{k=m}^{\infty} \frac{(\lambda/\mu)^k}{k!}\right]\left[\sum_{k=0}^{m-1} \frac{(\lambda/\mu)^k}{k!}\right] < \left(\frac{(m\rho)^m}{m!}\right)\left(\frac{1}{1-\rho}\right)\left(e^{\lambda/\mu} - \sum_{k=m}^{\infty} \frac{(\lambda/\mu)^k}{k!}\right)$$

$$\sum_{k=m}^{\infty} \frac{(\lambda/\mu)^k}{k!}e^{-\lambda/\mu} < \frac{\left(\frac{(m\rho)^m}{m!}\right)\left(\frac{1}{1-\rho}\right)}{\left[\sum_{k=0}^{m-1} \frac{(\lambda/\mu)^k}{k!} + \left(\frac{(m\rho)^m}{m!}\right)\left(\frac{1}{1-\rho}\right)\right]}$$

$$= C\left(m, \frac{\lambda}{\mu}\right)$$

(b) Since

$$1 - \frac{\lambda}{m\mu}\left[1 - B\left(m, \frac{\lambda}{\mu}\right)\right] = \frac{\left[1 - \frac{\lambda}{m\mu}\right]\left[\sum_{k=0}^{m-1}\frac{(\lambda/\mu)^k}{k!}\right] + \frac{(\lambda/\mu)^m}{m!}}{\sum_{k=0}^{m}\frac{(\lambda/\mu)^k}{k!}}$$

we have

$$\frac{B\left(m, \frac{\lambda}{\mu}\right)}{1 - \frac{\lambda}{m\mu}\left[1 - B\left(m, \frac{\lambda}{\mu}\right)\right]} = \frac{\frac{(\lambda/\mu)^m}{m!}}{\left[1 - \frac{\lambda}{m\mu}\right]\left[\sum_{k=0}^{m-1}\frac{(\lambda/\mu)^k}{k!}\right] + \frac{(\lambda/\mu)^m}{m!}}$$

$$= \frac{\left(\frac{(m\rho)^m}{m!}\right)\left(\frac{1}{1-\rho}\right)}{\left[\sum_{k=0}^{m-1}\frac{(\lambda/\mu)^k}{k!} + \left(\frac{(m\rho)^m}{m!}\right)\left(\frac{1}{1-\rho}\right)\right]}$$

$$= C\left(m, \frac{\lambda}{\mu}\right)$$

(c) We have

$$
B\left(m+1, \frac{\lambda}{\mu}\right) = \frac{\dfrac{(\lambda/\mu)^{m+1}}{(m+1)!}}{\displaystyle\sum_{k=0}^{m+1} \frac{(\lambda/\mu)^k}{k!}} = \frac{\left[\dfrac{(\lambda/\mu)^m}{m!}\right]\left(\dfrac{\lambda}{\mu}\right)}{(m+1)\displaystyle\sum_{k=0}^{m}\frac{(\lambda/\mu)^k}{k!} + \left[\dfrac{(\lambda/\mu)^m}{m!}\right]\left(\dfrac{\lambda}{\mu}\right)}
$$

$$
= \frac{\dfrac{\lambda}{\mu} B\left(m, \dfrac{\lambda}{\mu}\right)}{m+1+\dfrac{\lambda}{\mu} B\left(m, \dfrac{\lambda}{\mu}\right)} \qquad\qquad \square
$$

PROBLEM 3.10

Here we consider an M/M/1 queue in discrete time, where time is segmented into intervals of length q sec each. We assume that events can only occur at the ends of these discrete-time intervals. In particular, the probability of a single arrival at the end of such an interval is given by λq and the probability of no arrival at that point is $1 - \lambda q$ (thus at most one arrival may occur). Similarly, the departure process is such that if a customer is in service during an interval he will complete service at the end of that interval with probability $1 - \sigma$ or will require at least one more interval with probability σ.

(a) Derive the form for $a(t)$ and $b(x)$, the interarrival time and service time pdf's, respectively.

(b) Assuming FCFS, write down the equilibrium equations that govern the behavior of $p_k = P[k$ customers in system at the end of a discrete-time interval], where k includes any arrivals who have occurred at the end of this interval as well as any customers who are about to leave at this point.

(c) Solve for the expected value of the number of customers at these points.

SOLUTION

(a) In order that the interarrival time $\tilde{t} = nq$, we require no arrivals for $n-1$ consecutive intervals, followed by an arrival.

$$
P[\tilde{t} = nq] = \lambda q(1 - \lambda q)^{n-1} \qquad n = 1, 2, \ldots
$$

Similarly,

$$
P[\tilde{x} = nq] = (1 - \sigma)\sigma^{n-1} \qquad n = 1, 2, \ldots
$$

and so

$$a(t) = \sum_{n=1}^{\infty} \lambda q (1 - \lambda q)^{n-1} u_0(t - nq)$$

$$b(x) = \sum_{n=1}^{\infty} (1 - \sigma)\sigma^{n-1} u_0(x - nq)$$

(b) The number in system is a discrete-state discrete-time Markov chain, the equilibrium equations of which are

$$\lambda q p_0 = (1 - \lambda q)(1 - \sigma)p_1 \qquad k = 0$$

$$[\lambda q \sigma + (1 - \lambda q)(1 - \sigma)]p_1 = \lambda q p_0 + (1 - \lambda q)(1 - \sigma)p_2 \qquad k = 1$$

$$[\lambda q \sigma + (1 - \lambda q)(1 - \sigma)]p_k = \lambda q \sigma p_{k-1} + (1 - \lambda q)(1 - \sigma)p_{k+1} \qquad k \geq 2$$

(c) First we solve for p_k by iteration. We find $\rho = \bar{x}/\bar{t}$ as follows:

$$\bar{t} = \sum_{n=1}^{\infty} (nq)\lambda q(1 - \lambda q)^{n-1} = \frac{1}{\lambda}$$

$$\bar{x} = \sum_{n=1}^{\infty} (nq)(1 - \sigma)\sigma^{n-1} = \frac{q}{1 - \sigma}$$

and thus

$$\rho = \frac{\lambda q}{1 - \sigma}$$

Now for the iteration. We have

$$p_1 = \frac{\rho}{1 - \lambda q}p_0$$

$$p_2 = \frac{\rho^2 \sigma}{(1 - \lambda q)^2}p_0$$

and since

$$p_k = \frac{\lambda q \sigma}{(1 - \lambda q)(1 - \sigma)}p_{k-1} \qquad k \geq 3$$

our iteration yields

$$p_k = \left[\frac{\rho \sigma}{1 - \lambda q}\right]^k \frac{1}{\sigma}p_0 \qquad \text{for } k \geq 1$$

[Note: For stability (see Eq. (1.32)), $\rho\sigma/(1 - \lambda q) < 1$; that is, $\lambda q \sigma < (1 - \lambda q)(1 - \sigma)$. Thus $\lambda q \sigma < 1 - \lambda q - \sigma + \lambda q \sigma$ or $\lambda q < 1 - \sigma$; that is,

$\rho < 1$.] To find p_0, we use $\sum_{k=0}^{\infty} p_k = 1$ and obtain

$$p_0 + \sum_{k=1}^{\infty} \left[\frac{\rho\sigma}{1 - \lambda q}\right]^k \frac{1}{\sigma} p_0 = 1$$

$$p_0 \left(1 + \frac{1}{\sigma} \left[\frac{1}{1 - \dfrac{\rho\sigma}{1 - \lambda q}} - 1\right]\right) = 1$$

$$p_0 \left(1 + \frac{1}{\sigma} \frac{\rho\sigma}{1 - \lambda q - \rho\sigma}\right) = 1$$

Therefore

$$p_0 = 1 - \rho$$

$$p_k = \frac{1 - \rho}{\sigma} \left[\frac{\rho\sigma}{1 - \lambda q}\right]^k \qquad k \geq 1$$

Finally, the mean number in system at these points in time is

$$\overline{N} = \sum_{k=1}^{\infty} k p_k = \frac{1 - \rho}{\sigma} \frac{\rho\sigma}{1 - \lambda q} \sum_{k=1}^{\infty} k \left[\frac{\rho\sigma}{1 - \lambda q}\right]^{k-1}$$

$$= \frac{(1 - \rho)\rho}{1 - \lambda q} \frac{1}{\left(1 - \dfrac{\rho\sigma}{1 - \lambda q}\right)^2}$$

$$\overline{N} = \frac{\rho}{1 - \rho}(1 - \lambda q)$$

[$\overline{N}$ can also be found from the z-transform $P(z) = \sum_{k=0}^{\infty} p_k z^k$, which turns out to be

$$P(z) = (1 - \rho)\left(1 + \frac{\lambda q z}{(1 - \lambda q)(1 - \sigma) - \lambda q \sigma z}\right)$$

Then $\overline{N} = P^{(1)}(1)$.]   □

PROBLEM 3.11

Consider an M/M/1 system with "feedback"; by this we mean that when a customer departs from service he has probability σ of rejoining the tail of the queue after a random feedback time, which is exponentially distributed (with mean $1/\gamma$ sec); on the other hand, with probability $1 - \sigma$ he will depart forever after completing service. It is clear that a customer may return many times to the tail of the queue before making his eventual final departure (see figure). Let p_{kj} be the equilibrium probability that there are k customers in the "system" (i.e., in the queue and the service facility) and that there are j customers in the process of returning to the system.

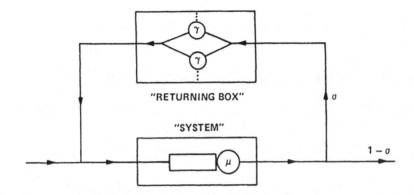

(a) Write down the set of difference equations for the equilibrium probabilities p_{kj}.

(b) Defining the double z-transform

$$P(z_1, z_2) = \sum_{k=0}^{\infty} \sum_{j=0}^{\infty} p_{kj} z_1^k z_2^j$$

show that

$$\gamma(z_2 - z_1)\frac{\partial P(z_1, z_2)}{\partial z_2} + \left\{ \lambda(1 - z_1) + \mu \left[1 - \frac{1-\sigma}{z_1} - \sigma\frac{z_2}{z_1} \right] \right\} P(z_1, z_2)$$

$$= \mu \left[1 - \frac{1-\sigma}{z_1} - \sigma\frac{z_2}{z_1} \right] P(0, z_2)$$

(c) By taking advantage of the moment-generating properties of our z-transforms, show that the mean number in the "system" (queue plus server) is given by $\rho/(1 - \rho)$ and that the mean number returning to the tail of the queue is given by $\mu\sigma\rho/\gamma$, where $\rho = \lambda/(1 - \sigma)\mu$.

SOLUTION

(a) **Case (1):** $k = 0$, $j = 0$

$$\lambda p_{00} = \mu(1 - \sigma)p_{10}$$

Case (2): $k = 0$, $j > 0$

$$(\lambda + j\gamma)p_{0j} = \mu(1 - \sigma)p_{1j} + \mu\sigma p_{1,j-1}$$

Case (3): $k > 0$, $j = 0$

$$(\lambda + \mu)p_{k0} = \lambda p_{k-1,0} + \mu(1 - \sigma)p_{k+1,0} + \gamma p_{k-1,1}$$

Case (4): $k > 0$, $j > 0$

$$(\lambda + \mu + j\gamma)p_{kj} = \lambda p_{k-1,j} + \mu(1 - \sigma)p_{k+1,j}$$
$$+ \mu\sigma p_{k+1,j-1} + (j + 1)\gamma p_{k-1,j+1}$$

(b) We multiply the (k, j) equation by $z_1^k z_2^j$ and sum, case by case:

Case (1) yields

$$\lambda p_{00} = \mu(1 - \sigma)p_{10}$$

Case (2) yields

$$\sum_{j=1}^{\infty}(\lambda + j\gamma)p_{0j}z_2^j = \sum_{j=1}^{\infty}\mu(1 - \sigma)p_{1j}z_2^j + \sum_{j=1}^{\infty}\mu\sigma p_{1,j-1}z_2^j$$

Case (3) yields

$$\sum_{k=1}^{\infty}(\lambda + \mu)p_{k0}z_1^k = \sum_{k=1}^{\infty}\lambda p_{k-1,0}z_1^k + \sum_{k=1}^{\infty}\mu(1 - \sigma)p_{k+1,0}z_1^k + \sum_{k=1}^{\infty}\gamma p_{k-1,1}z_1^k$$

Case (4) yields

$$\sum_{k=1}^{\infty}\sum_{j=1}^{\infty}(\lambda + \mu + j\gamma)p_{kj}z_1^k z_2^j$$

$$= \sum_{k=1}^{\infty}\sum_{j=1}^{\infty}\lambda p_{k-1,j}z_1^k z_2^j + \sum_{k=1}^{\infty}\sum_{j=1}^{\infty}\mu(1 - \sigma)p_{k+1,j}z_1^k z_2^j$$

$$+ \sum_{k=1}^{\infty}\sum_{j=1}^{\infty}\mu\sigma p_{k+1,j-1}z_1^k z_2^j + \sum_{k=1}^{\infty}\sum_{j=1}^{\infty}(j + 1)\gamma p_{k-1,j+1}z_1^k z_2^j$$

Summing these four equations gives

$$\lambda P(z_1, z_2) + \mu\left[P(z_1, z_2) - \sum_{j=0}^{\infty}p_{0j}z_2^j\right] + \gamma z_2\frac{\partial}{\partial z_2}P(z_1, z_2)$$

$$= \lambda z_1 P(z_1, z_2) + \frac{\mu(1 - \sigma)}{z_1}\left[P(z_1, z_2) - \sum_{j=0}^{\infty}p_{0j}z_2^j\right]$$

$$+ \mu\sigma\frac{z_2}{z_1}\left[P(z_1, z_2) - \sum_{j=0}^{\infty}p_{0j}z_2^j\right] + \gamma z_1\frac{\partial}{\partial z_2}P(z_1, z_2)$$

Noting that

$$P(z_1, z_2) = \sum_{j=0}^{\infty} p_{0j} z_2^j + \sum_{k=1}^{\infty} \sum_{j=0}^{\infty} p_{kj} z_1^k z_2^j$$

and thus

$$P(0, z_2) = \sum_{j=0}^{\infty} p_{0j} z_2^j$$

we finally have the required equation:

$$\gamma(z_2 - z_1) \frac{\partial P(z_1, z_2)}{\partial z_2} + \left\{ \lambda(1 - z_1) + \mu \left[1 - \frac{1 - \sigma}{z_1} - \sigma \frac{z_2}{z_1} \right] \right\} P(z_1, z_2)$$

$$= \mu \left[1 - \frac{1 - \sigma}{z_1} - \sigma \frac{z_2}{z_1} \right] P(0, z_2)$$

(c) The mean number of customers in the "returning box" is

$$\overline{N}_2 = \sum_{j=1}^{\infty} j \left(\sum_{k=0}^{\infty} p_{kj} \right)$$

which may also be written as

$$\overline{N}_2 = \frac{\partial}{\partial z_2} P(z_1, z_2) \bigg|_{z_1 = z_2 = 1}$$

From this, we find $\overline{N}_2$ as follows. First, substitute $z_1 = 1$ in the equation derived in part (b) to give

$$\gamma(z_2 - 1) \frac{\partial}{\partial z_2} P(1, z_2) + \mu\sigma(1 - z_2) P(1, z_2) = \mu\sigma(1 - z_2) P(0, z_2)$$

Now, for $z_2 \neq 1$,

$$\gamma \frac{\partial}{\partial z_2} P(1, z_2) = \mu\sigma [P(1, z_2) - P(0, z_2)]$$

Letting $z_2 \to 1$ and using analyticity we obtain

$$\gamma \overline{N}_2 = \mu\sigma [P(1, 1) - P(0, 1)]$$

Clearly $P(1, 1) = 1$. To find $P(0, 1)$, we put $z_1 = z_2 = z$ in the equation from part (b):

$$\left[\lambda(1 - z) + \mu(1 - \sigma) \left(1 - \frac{1}{z} \right) \right] P(z, z) = \mu(1 - \sigma) \left(1 - \frac{1}{z} \right) P(0, z)$$

or

$$P(z,z)[\mu(1 - \sigma) - \lambda z] = \mu(1 - \sigma)P(0,z)$$

Letting $z \to 1$, we get

$$P(1,1)[\mu(1 - \sigma) - \lambda] = \mu(1 - \sigma)P(0,1)$$

and so

$$P(0,1) = 1 - \frac{\lambda}{\mu(1 - \sigma)} = 1 - \rho$$

Therefore

$$\overline{N}_2 = \frac{\mu\sigma\rho}{\gamma}$$

[Note: The arrival rate, say, η, to the "system" satisfies $\eta = \lambda + \eta\sigma$ or $\eta = \lambda/(1 - \sigma)$. Therefore $\rho = \eta/\mu = \lambda/\mu(1 - \sigma)$.]

We observe that the mean number of customers in the "system" is

$$\overline{N}_1 = \sum_{k=1}^{\infty} k \left(\sum_{j=0}^{\infty} p_{kj} \right)$$

In order to find $\overline{N}_1$, we first define

$$Q(z) = P(z,z) = \sum_{k=0}^{\infty} \sum_{j=0}^{\infty} p_{kj} z^{k+j}$$

Therefore,

$$\frac{dQ(z)}{dz}\bigg|_{z=1} = \sum_{k=0}^{\infty} \sum_{j=0}^{\infty} (k + j)p_{kj} = \overline{N}_1 + \overline{N}_2$$

We already know that

$$P(z,z)[\mu(1 - \sigma) - \lambda z] = \mu(1 - \sigma)P(0,z)$$

and so

$$Q(z)(1 - \rho z) = P(0,z)$$

Differentiating with respect to z gives

$$\frac{dQ(z)}{dz}(1 - \rho z) - \rho Q(z) = \frac{\partial}{\partial z}P(0,z)$$

Letting $z \rightarrow 1$ (use $Q(1) = 1$) yields

$$\left(\overline{N}_1 + \overline{N}_2\right)(1 - \rho) = \rho + \left.\frac{\partial P(0, z)}{\partial z}\right|_{z=1}$$

$$\overline{N}_1 + \overline{N}_2 = \frac{\rho}{1 - \rho} + \left(\frac{1}{1 - \rho}\right)\left.\frac{\partial P(0, z)}{\partial z}\right|_{z=1}$$

Now

$$\left(\frac{1}{1 - \rho}\right)\left.\frac{\partial P(0, z)}{\partial z}\right|_{z=1} = \frac{\left.\dfrac{\partial P(0, z)}{\partial z}\right|_{z=1}}{P(0, 1)}$$

and from the expression we derived in (b) for $P(0, z_2)$, we see that this represents the mean number of customers in the "returning box" conditioned on there being no one in the "system." Assuming this is independent of the state of the "system," it is just $\overline{N}_2$. Thus

$$\overline{N}_1 = \frac{\rho}{1 - \rho} \qquad \square$$

PROBLEM 3.12

Consider a "cyclic queue" in which M customers circulate around through two queueing facilities as shown below.

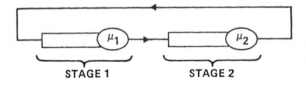

STAGE 1   STAGE 2

Both servers are of the exponential type with rates μ_1 and μ_2, respectively. Let

$$p_k = P[k \text{ customers in stage 1 and } M - k \text{ in stage 2}]$$

(a) Draw the state-transition-rate diagram.
(b) Write down the relationship among $\{p_k\}$.
(c) Find

$$P(z) = \sum_{k=0}^{M} p_k z^{-k}$$

(d) Find p_k.

SOLUTION

(a) The state-transition-rate diagram is

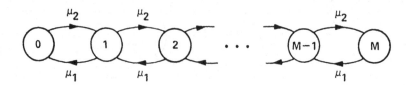

[It turns out that this is exactly the state-transition-rate diagram of the finite storage system M/M/1/K studied in part (iv) of Problem 3.8, with $K = M$, $\lambda = \mu_2$, and $\mu = \mu_1$. Therefore we need proceed no further to find p_k. However, in this problem, we seek the balance equations and the z-transform $P(z)$.]

(b) The balance equations are

$$\mu_2 p_0 = \mu_1 p_1 \qquad k = 0$$

$$(\mu_1 + \mu_2)p_k = \mu_2 p_{k-1} + \mu_1 p_{k+1} \qquad 1 \le k \le M - 1$$

$$\mu_1 p_M = \mu_2 p_{M-1} \qquad k = M$$

(c) Multiplying the kth equation ($0 \le k \le M$) by z^k and summing,

$$\mu_1 \sum_{k=1}^{M} p_k z^k + \mu_2 \sum_{k=0}^{M-1} p_k z^k = \mu_1 \sum_{k=0}^{M-1} p_{k+1} z^k + \mu_2 \sum_{k=1}^{M} p_{k-1} z^k$$

and so

$$\mu_1 [P(z) - p_0] + \mu_2 [P(z) - p_M z^M] = \frac{\mu_1}{z} [P(z) - p_0] + \mu_2 z [P(z) - p_M z^M]$$

Therefore

$$P(z) = \frac{p_0 - \dfrac{\mu_2}{\mu_1} z p_M z^M}{1 - \dfrac{\mu_2}{\mu_1} z}$$

The polynomial $P(z)$ is analytic everywhere in the finite z-plane, and so the root $z = \mu_1/\mu_2$ of the denominator must also be a root of the numerator. Thus

$$p_0 - \frac{\mu_2}{\mu_1} \left(\frac{\mu_1}{\mu_2}\right) p_M \left(\frac{\mu_1}{\mu_2}\right)^M = 0$$

or

$$p_M = p_0 \left(\frac{\mu_2}{\mu_1}\right)^M$$

Therefore

$$P(z) = \frac{p_0 - p_0 \left(\frac{\mu_2}{\mu_1}\right)^{M+1} z^{M+1}}{1 - \frac{\mu_2}{\mu_1} z}$$

or

$$P(z) = p_0 \sum_{k=0}^{M} \left(\frac{\mu_2}{\mu_1} z\right)^k$$

Now for p_0:

$$1 = P(1) = p_0 \sum_{k=0}^{M} \left(\frac{\mu_2}{\mu_1}\right)^k$$

So

$$P(z) = \frac{\sum_{k=0}^{M} \left(\frac{\mu_2}{\mu_1}\right)^k z^k}{\sum_{k=0}^{M} \left(\frac{\mu_2}{\mu_1}\right)^k}$$

(d) By inspection of the power series in (c), we get

$$p_k = \frac{\left(\frac{\mu_2}{\mu_1}\right)^k}{\sum_{k=0}^{M} \left(\frac{\mu_2}{\mu_1}\right)^k} \qquad 0 \le k \le M$$

which may also be expressed as

$$p_k = \frac{1 - \frac{\mu_2}{\mu_1}}{1 - \left(\frac{\mu_2}{\mu_1}\right)^{M+1}} \left(\frac{\mu_2}{\mu_1}\right)^k \qquad 0 \le k \le M \qquad \square$$

PROBLEM 3.13

Consider an M/M/1 queue with parameters λ and μ. A customer in the queue will defect (depart without service) with probability $\alpha \Delta t + o(\Delta t)$ in any interval of duration Δt.

(a) Draw the state-transition-rate diagram.
(b) Express p_{k+1} in terms of p_k.
(c) For $\alpha = \mu$, solve for p_k ($k = 0, 1, 2, \ldots$).

SOLUTION

(a) The state-transition-rate diagram is

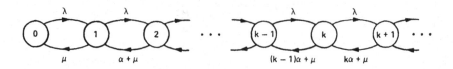

(b) The balance equations are

$$\lambda p_0 = \mu p_1 \qquad k = 0$$

$$\left[\lambda + (k - 1)\alpha + \mu\right]p_k = \lambda p_{k-1} + (k\alpha + \mu)p_{k+1} \qquad k \geq 1$$

By recursion we find

$$(k\alpha + \mu)p_{k+1} = \lambda p_k$$

(c) First observe from part (b) that, for $k \geq 1$, we have $\mu_k = (k - 1)\alpha + \mu$. Thus $\mu_k = k\mu$ for $\alpha = \mu$. Note that λ_k and μ_k for this system ($\alpha = \mu$) are exactly the same as for M/M/∞. Therefore the solution for p_k must be the same (although other system parameters such as waiting time and queue size will be different). Then Eq. (1.89) gives

$$p_k = \frac{\left(\dfrac{\lambda}{\mu}\right)^k}{k!}e^{-\frac{\lambda}{\mu}} \qquad k = 0, 1, 2, \ldots \qquad \square$$

PROBLEM 3.14

Let us elaborate on the M/M/1/K system of Problem 3.8.

(a) Evaluate p_k when $\lambda = \mu$.
(b) Find $\bar{N}$ for $\lambda \neq \mu$ and for $\lambda = \mu$.
(c) Find T by carefully solving for the average arrival rate to the system.

SOLUTION

(a) Equation (1.63) gives

$$p_k = p_0 \left(\frac{\lambda}{\mu}\right)^k \qquad k \leq K$$

$$p_k = 0 \qquad k > K$$

Thus for $\lambda = \mu$ we find

$$p_k = p_0 \qquad \text{for } k \leq K$$

Since $\sum_{k=0}^{K} p_k = 1$, then

$$p_k = \frac{1}{K+1} \qquad 0 \leq k \leq K$$

(b) **(i)** For $\lambda = \mu$,

$$\overline{N} = \sum_{k=1}^{K} k p_k = \sum_{k=1}^{K} k \frac{1}{K+1} = \frac{1}{K+1} \sum_{k=1}^{K} k = \frac{1}{K+1} \cdot \frac{K(K+1)}{2}$$

$$\overline{N} = \frac{K}{2}$$

(ii) For $\lambda \neq \mu$,

$$\overline{N} = \sum_{k=1}^{K} k p_k = \sum_{k=1}^{K} k p_0 \left(\frac{\lambda}{\mu}\right)^k = p_0 \left(\frac{\lambda}{\mu}\right) \frac{d}{d\left(\frac{\lambda}{\mu}\right)} \left[\sum_{k=0}^{K} \left(\frac{\lambda}{\mu}\right)^k\right]$$

Using Eq. (1.63), we find that

$$p_0 = \frac{1 - \dfrac{\lambda}{\mu}}{1 - \left(\dfrac{\lambda}{\mu}\right)^{K+1}}$$

So

$$\overline{N} = \frac{1 - \dfrac{\lambda}{\mu}}{1 - \left(\dfrac{\lambda}{\mu}\right)^{K+1}} \left(\frac{\lambda}{\mu}\right) \left[\frac{1 - (K+1)\left(\dfrac{\lambda}{\mu}\right)^K + K\left(\dfrac{\lambda}{\mu}\right)^{K+1}}{\left(1 - \dfrac{\lambda}{\mu}\right)^2}\right]$$

or

$$\overline{N} = \left(\frac{\lambda}{\mu}\right) \left[\frac{1 - (K+1)\left(\dfrac{\lambda}{\mu}\right)^K + K\left(\dfrac{\lambda}{\mu}\right)^{K+1}}{\left(1 - \left(\dfrac{\lambda}{\mu}\right)^{K+1}\right)\left(1 - \dfrac{\lambda}{\mu}\right)}\right]$$

(c) **(i)** For $\lambda = \mu$, $p_k = 1/(K+1)$, $0 \leq k \leq K$ from part (a). The average arrival rate $\overline{\lambda}$ is

$$\overline{\lambda} = \sum_{k=0}^{K} \lambda_k p_k = \lambda \sum_{k=0}^{K-1} \frac{1}{K+1} \qquad (\lambda_K = 0)$$

$$\overline{\lambda} = \lambda \frac{K}{K+1}$$

By Little's result, and also part (b)

$$T = \frac{\overline{N}}{\overline{\lambda}} = \frac{K/2}{\lambda K/(K+1)} = \frac{K+1}{2\lambda}$$

(ii) For $\lambda \neq \mu$, from Eq. (1.63)

$$\overline{\lambda} = \sum_{k=0}^{K} \lambda_k p_k = \lambda \sum_{k=0}^{K-1} \frac{1 - \dfrac{\lambda}{\mu}}{1 - \left(\dfrac{\lambda}{\mu}\right)^{K+1}} \left(\frac{\lambda}{\mu}\right)^k$$

$$\overline{\lambda} = \lambda \cdot \frac{1 - \left(\dfrac{\lambda}{\mu}\right)^{K}}{1 - \left(\dfrac{\lambda}{\mu}\right)^{K+1}}$$

Little's result and part (b) give

$$T = \frac{\overline{N}}{\overline{\lambda}}$$

$$= \left(\frac{\lambda}{\mu}\right) \left[\frac{1 - (K+1)\left(\dfrac{\lambda}{\mu}\right)^{K} + K\left(\dfrac{\lambda}{\mu}\right)^{K+1}}{\left(1 - \left(\dfrac{\lambda}{\mu}\right)^{K+1}\right)\left(1 - \dfrac{\lambda}{\mu}\right)} \right] \left[\frac{1 - \left(\dfrac{\lambda}{\mu}\right)^{K+1}}{\lambda\left(1 - \left(\dfrac{\lambda}{\mu}\right)^{K}\right)} \right]$$

$$T = \frac{1}{\mu} \left[\frac{1 - (K+1)\left(\dfrac{\lambda}{\mu}\right)^{K} + K\left(\dfrac{\lambda}{\mu}\right)^{K+1}}{\left(1 - \left(\dfrac{\lambda}{\mu}\right)^{K}\right)\left(1 - \dfrac{\lambda}{\mu}\right)} \right] \qquad \square$$

CHAPTER 4

MARKOVIAN QUEUES

PROBLEM 4.1

Consider the Markovian queueing system shown below. Branch labels are birth and death rates. Node labels give the number of customers in the system.

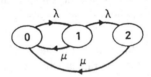

(a) Solve for p_k.

(b) Find the average number in the system.

(c) For $\lambda = \mu$, what values do we get for parts (a) and (b)? Try to interpret these results.

(d) Write down the transition rate matrix $\mathbf{Q}$ for this problem and give the matrix equation relating $\mathbf{Q}$ to the probabilities found in part (a).

SOLUTION

(a) Using the flow conservation law for states 0 and 2 and the conservation of probability, we get the following three independent equations:

$$\lambda p_0 = \mu p_1 + \mu p_2$$

$$\mu p_2 = \lambda p_1$$

$$p_0 + p_1 + p_2 = 1$$

97

Solving this gives

$$p_0 = \frac{\mu}{\lambda + \mu}$$

$$p_1 = \frac{\lambda\mu}{(\lambda + \mu)^2}$$

$$p_2 = \frac{\lambda^2}{(\lambda + \mu)^2}$$

(b) We have

$$\overline{N} = 0 \cdot p_0 + 1 \cdot p_1 + 2 \cdot p_2 = \frac{\lambda\mu + 2\lambda^2}{(\lambda + \mu)^2}$$

$$\overline{N} = \frac{\lambda(2\lambda + \mu)}{(\lambda + \mu)^2}$$

(c) If $\lambda = \mu$, the results in parts (a) and (b) become

$$p_0 = \tfrac{1}{2}, \; p_1 = p_2 = \tfrac{1}{4}, \; \overline{N} = \tfrac{3}{4}$$

To interpret these results, consider a cycle from state 0 back to state 0. The rate out of state 0 is λ ($= \mu$), which puts the system into state 1. The rate out of state 1 is $\lambda + \mu = 2\mu$, and so the fraction of time spent in state 1 must be half that spent in state 0. From state 1 we arrive at state 2 with probability $\tfrac{1}{2}$ (or return directly to state 0 with probability $\tfrac{1}{2}$) and depart state 2 at rate μ; therefore we spend as much time, on the average, in state 2, (i.e., $\tfrac{1}{2} \cdot (1/\mu)$) as in state 1 (i.e., $1/2\mu$).

(d) Equation (1.53) implies that $-q_{ii}$ is the rate at which the system departs from state i, while q_{ij} ($i \neq j$) is the rate at which it moves from state i to state j. Thus

$$Q = \begin{bmatrix} -\lambda & \lambda & 0 \\ \mu & -(\mu + \lambda) & \lambda \\ \mu & 0 & -\mu \end{bmatrix}$$

From Eq. (1.56) we have directly that

$$\pi Q = 0 \qquad (\pi = p = [p_0, p_1, p_2]) \qquad \Box$$

PROBLEM 4.2

Consider an $E_k/E_n/1$ queueing system where *no* queue is permitted to form. A customer who arrives to find the service facility busy is "lost" (he departs with no service). Let (i, j) be the system state in which the "arriving" customer is in the ith arrival stage and the customer in service is in the jth service stage (note that there is always some customer in the arrival mechanism and that if there is no customer in the service facility, then we let $j = 0$). Let $1/k\lambda$ be the average time spent in any arrival stage and $1/n\mu$ be the average time spent in any service stage.

(a) Draw the state diagram showing all the transition *rates*.

(b) Write down the equilibrium equation for (i, j) where $1 < i < k, 1 < j \leq n$.

SOLUTION

(a) The state-transition-rate diagram is

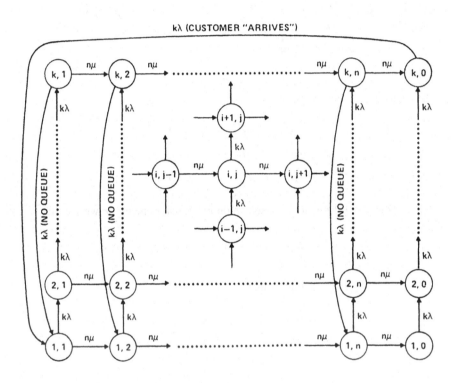

(b) Using Flow Out = Flow In, we obtain

$$(k\lambda + n\mu)p_{ij} = k\lambda p_{i-1,j} + n\mu p_{i,j-1} \qquad \text{for } 1 < i < k, \ 1 < j \leq n \quad \square$$

PROBLEM 4.3

Consider an M/E$_r$/1 system in which *no* queue is allowed to form. Let j be the number of stages of service left in the system and let P_j be the equilibrium probability of being in state (i, j).

(a) Find $P_j, j = 0, 1, \ldots, r$.

(b) Find the probability of a busy system.

SOLUTION

The state-transition-rate diagram is

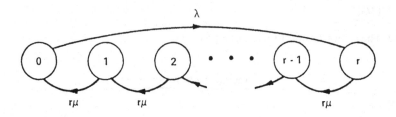

(a) The flow equations are

$$\lambda P_0 = r\mu P_1 \qquad j = 0$$

$$r\mu P_j = r\mu P_{j+1} \qquad 1 \le j \le r - 1$$

$$r\mu P_r = \lambda P_0 \qquad j = r$$

Of these $r + 1$ equations, one is redundant; using the first r we see that

$$\frac{\lambda}{r\mu} P_0 = P_1 = P_2 = \cdots = P_{r-1} = P_r$$

Also $\sum_{j=0}^{r} P_j = 1$ implies that

$$P_0 + \sum_{j=1}^{r} \frac{\lambda}{r\mu} P_0 = 1$$

Thus

$$P_0 = \frac{\mu}{\lambda + \mu}$$

and therefore

$$P_j := \frac{\lambda}{r(\lambda + \mu)} \qquad 1 \le j \le r$$

(b) We have

$$P[\text{busy system}] = 1 - P_0 = 1 - \frac{\mu}{\lambda + \mu}$$

$$P[\text{busy system}] = \frac{\lambda}{\lambda + \mu} \qquad\qquad \square$$

PROBLEM 4.4

Consider an $M/H_2/1$ system in which *no* queue is allowed to form. Service is of the hyperexponential type with $\mu_1 = 2\mu\alpha_1$ and $\mu_2 = 2\mu(1 - \alpha_1)$.

  (a) Solve for the equilibrium probability of an empty system.
  (b) Find the probability that stage 1 is occupied.
  (c) Find the probability of a busy system.

SOLUTION

Let 1_i represent the state when there is one customer in the system and that customer is in stage i. The state diagram for this system is as follows:

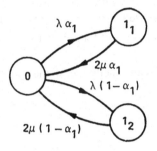

As usual, we have two independent flow equations and the conservation of probability:

$$\lambda p_0 = 2\mu\alpha_1 p_{1_1} + 2\mu(1 - \alpha_1)p_{1_2}$$

$$\lambda\alpha_1 p_0 = 2\mu\alpha_1 p_{1_1}$$

$$p_0 + p_{1_1} + p_{1_2} = 1$$

Thus

$$p_0 = \frac{\mu}{\lambda + \mu}$$

$$p_{1_1} = p_{1_2} = \frac{\lambda}{2(\lambda + \mu)}$$

  (a) The probability of an empty system is

$$P[\text{empty system}] = p_0 = \frac{\mu}{\lambda + \mu}$$

  (b) The probability that stage 1 is busy is

$$P[\text{stage 1 busy}] = p_{1_1} = \frac{\lambda}{2(\lambda + \mu)}$$

(c) The probability of a busy system is

$$P[\text{busy system}] = 1 - p_0 = p_{1_1} + p_{1_2} = \frac{\lambda}{\lambda + \mu} \qquad \square$$

PROBLEM 4.5

Consider an M/M/1 system with parameters λ and μ in which exactly two customers arrive at each arrival instant.

(a) Draw the state-transition-rate diagram.
(b) By inspection, write down the equilibrium equations for p_k $(k = 0, 1, 2, \ldots)$.
(c) Let $\rho = 2\lambda/\mu$. Express $P(z)$ in terms of ρ and z.
(d) Find $P(z)$ by using the bulk arrival result given in Eq. (1.82).
(e) Find the mean and variance of the number of customers in the system from $P(z)$.
(f) Repeat parts (a)–(e) with exactly r customers arriving at each arrival instant (and $\rho = r\lambda/\mu$).

SOLUTION

(a) The state-transition-rate diagram is as follows:

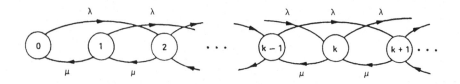

(b) The equilibrium equations are

$$\lambda p_0 = \mu p_1 \qquad k = 0$$
$$(\lambda + \mu)p_1 = \mu p_2 \qquad k = 1$$
$$(\lambda + \mu)p_k = \lambda p_{k-2} + \mu p_{k+1} \qquad k \geq 2$$

(c) Multiply the kth equation by z^k and sum for $k \geq 0$. This gives

$$\lambda \sum_{k=0}^{\infty} p_k z^k + \mu \sum_{k=1}^{\infty} p_k z^k = \lambda \sum_{k=2}^{\infty} p_{k-2} z^k + \mu \sum_{k=0}^{\infty} p_{k+1} z^k$$

$$\lambda P(z) + \mu[P(z) - p_0] = \lambda z^2 P(z) + \frac{\mu}{z}[P(z) - p_0]$$

$$P(z) = \frac{\mu p_0 \left(1 - \frac{1}{z}\right)}{\lambda(1 - z^2) + \mu \left(1 - \frac{1}{z}\right)} = \frac{\mu p_0}{\mu - \lambda z(z + 1)}$$

(Note that the average arrival rate $\bar{\lambda} = 2\lambda$, and so $\rho = \bar{\lambda}\bar{x} = 2\lambda/\mu$.) Thus

$$P(z) = \frac{2p_0}{2 - \rho z(z + 1)}$$

Since $P(1) = 1 = 2p_0/(2 - 2\rho)$ we have $p_0 = 1 - \rho$. Hence

$$P(z) = \frac{2(1 - \rho)}{2 - \rho z(z + 1)}$$

(d) By Eq. (1.82),

$$P(z) = \frac{\mu(1 - \rho)(1 - z)}{\mu(1 - z) - \lambda z[1 - G(z)]}$$

In the system under consideration, bulks have constant size 2. Thus $G(z) = z^2$ (and $\rho = \lambda G^{(1)}(1)/\mu = 2\lambda/\mu$). Therefore

$$P(z) = \frac{\mu(1 - \rho)(1 - z)}{\mu(1 - z) - \lambda z(1 - z^2)}$$

This simplifies as before to

$$P(z) = \frac{2(1 - \rho)}{2 - \rho z(z + 1)}$$

(e) The mean and variance of the number of customers may be found from the first and second derivatives of $P(z)$. We find that

$$\frac{dP(z)}{dz} = \frac{2(1 - \rho)\rho(2z + 1)}{[2 - \rho z(z + 1)]^2}$$

$$\overline{N} = \frac{dP(z)}{dz}\bigg|_{z=1} = \frac{2(1 - \rho)\rho(3)}{(2 - 2\rho)^2}$$

$$\overline{N} = \frac{3}{2}\frac{\rho}{1 - \rho}$$

After simplification, the second derivative is

$$\frac{d^2P(z)}{dz^2} = 4(1 - \rho)\rho \left[\frac{[2 - \rho z(z + 1)] + \rho(2z + 1)^2}{[2 - \rho z(z + 1)]^3}\right]$$

$$\overline{N^2} - \overline{N} = \frac{d^2P(z)}{dz^2}\bigg|_{z=1} = 4(1 - \rho)\rho \left[\frac{2 - 2\rho + 9\rho}{(2 - 2\rho)^3}\right]$$

$$= \frac{\rho}{2(1 - \rho)^2}(2 + 7\rho)$$

By definition, we may find the variance of N as

$$\sigma_N^2 = \overline{N^2} - (\overline{N})^2 = (\overline{N^2} - \overline{N}) + \overline{N} - (\overline{N})^2$$

$$= \frac{\rho}{2(1-\rho)^2}(2 + 7\rho) + \frac{3}{2}\frac{\rho}{1-\rho} - \frac{9}{4}\frac{\rho^2}{(1-\rho)^2}$$

$$\sigma_N^2 = \frac{\rho(10 - \rho)}{4(1-\rho)^2}$$

(f) The state-transition-rate diagram is

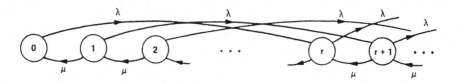

The equilibrium equations for p_k are

$$\lambda p_0 = \mu p_1 \qquad k = 0$$

$$(\lambda + \mu)p_k = \mu p_{k+1} \qquad 1 \le k \le r - 1$$

$$(\lambda + \mu)p_k = \lambda p_{k-r} + \mu p_{k+1} \qquad k \ge r$$

Multiply the kth equation by z^k and sum:

$$\lambda \sum_{k=0}^{\infty} p_k z^k + \mu \sum_{k=1}^{\infty} p_k z^k = \lambda \sum_{k=r}^{\infty} p_{k-r} z^k + \mu \sum_{k=0}^{\infty} p_{k+1} z^k$$

$$\lambda P(z) + \mu[P(z) - p_0] = \lambda z^r P(z) + \frac{\mu}{z}[P(z) - p_0]$$

$$P(z) = \frac{\mu p_0(z - 1)}{\mu(z - 1) - \lambda z(z^r - 1)}$$

$$P(z) = \frac{\mu p_0}{\mu - \lambda z \sum_{k=0}^{r-1} z^k}$$

As $\rho = r\lambda/\mu$ ($\overline{\lambda} = r\lambda$ and so $\rho = \overline{\lambda}\overline{x} = r\lambda/\mu$), we may write

$$P(z) = \frac{r p_0}{r - \rho \sum_{k=1}^{r} z^k}$$

Also $P(1) = 1 = r p_0/(r - r\rho)$ implies that $p_0 = 1 - \rho$. Thus

$$P(z) = \frac{r(1 - \rho)}{r - \rho \sum_{k=1}^{r} z^k}$$

To see this in another way, for the bulk arrival system with constant bulk size r, we have $G(z) = z^r$. Substituting this into Eq. (1.82) and simplifying gives as before

$$P(z) = \frac{r(1 - \rho)}{r - \rho \sum_{k=1}^{r} z^k}$$

To find $\overline{N}$ we note that

$$\frac{dP(z)}{dz} = r(1 - \rho)\rho \frac{\sum_{k=1}^{r} kz^{k-1}}{\left(r - \rho \sum_{k=1}^{r} z^k\right)^2}$$

so that

$$\overline{N} = \frac{dP(z)}{dz}\bigg|_{z=1} = r(1 - \rho)\rho \frac{r(r + 1)/2}{(r - r\rho)^2}$$

Thus

$$\overline{N} = \frac{r + 1}{2} \frac{\rho}{1 - \rho}$$

To find σ_N^2 we first obtain

$$\overline{N^2} - \overline{N} = \frac{d^2P(z)}{dz^2}\bigg|_{z=1} = r(1-\rho)\rho \left[\frac{(r - r\rho)\sum_{k=1}^{r} k(k-1) + 2\rho\left(\sum_{k=1}^{r} k\right)^2}{(r - r\rho)^3}\right]$$

Now recall that

$$\sum_{k=1}^{r} k = \frac{r(r + 1)}{2} \quad \text{and} \quad \sum_{k=1}^{r} k^2 = \frac{r(r + 1)(2r + 1)}{6}$$

Therefore

$$\sum_{k=1}^{r} k(k - 1) = \frac{(r - 1)r(r + 1)}{3}$$

and

$$\overline{N^2} - \overline{N} = r(1 - \rho)\rho \left[\frac{r(1 - \rho)\dfrac{(r - 1)r(r + 1)}{3} + 2\rho\left(\dfrac{r(r + 1)}{2}\right)^2}{[r(1 - \rho)]^3}\right]$$

$$= \frac{(r + 1)\rho}{6(1 - \rho)^2}(2r - 2 + \rho r + 5\rho)$$

and so

$$\sigma_N^2 = (\overline{N^2} - \overline{N}) + \overline{N} - (\overline{N})^2$$

$$= \frac{(r + 1)\rho}{6(1 - \rho)^2}(2r - 2 + \rho r + 5\rho) + \frac{(r + 1)\rho}{2(1 - \rho)} - \frac{(r + 1)^2\rho^2}{4(1 - \rho)^2}$$

$$\sigma_N^2 = \frac{(r + 1)\rho}{12(1 - \rho)^2}(4r + 2 - \rho r + \rho) \qquad \qquad \square$$

PROBLEM 4.6

Consider an M/M/1 queueing system with parameters λ and μ. At each of the arrival instants one new customer will enter the system with probability $\frac{1}{2}$ or two new customers will enter simultaneously with probability $\frac{1}{2}$.

(a) Draw the state-transition-rate diagram for this system.

(b) By inspection, write down the equilibrium equations for p_k.

(c) Find $P(z)$ and also evaluate any constants in this expression so that $P(z)$ is given in terms only of λ and μ. If possible eliminate any common factors in the numerator and denominator of this expression [this makes life simpler for you in part (d)].

(d) From part (c) find the expected number of customers in the system.

(e) Repeat part (c) using the bulk arrival results in Section 1.4 directly.

SOLUTION

(a) The state-transition-rate diagram is

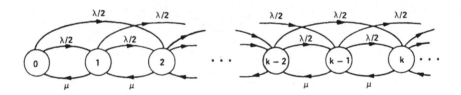

(b) The equilibrium equations are

$$\lambda p_0 = \mu p_1 \qquad k = 0$$

$$(\lambda + \mu)p_1 = \frac{\lambda}{2}p_0 + \mu p_2 \qquad k = 1$$

$$(\lambda + \mu)p_k = \frac{\lambda}{2}p_{k-2} + \frac{\lambda}{2}p_{k-1} + \mu p_{k+1} \qquad k \geq 2$$

(c) Multiply the kth equation by z^k and sum

$$\lambda \sum_{k=0}^{\infty} p_k z^k + \mu \sum_{k=1}^{\infty} p_k z^k = \frac{\lambda}{2} \sum_{k=2}^{\infty} p_{k-2} z^k + \frac{\lambda}{2} \sum_{k=1}^{\infty} p_{k-1} z^k + \mu \sum_{k=0}^{\infty} p_{k+1} z^k$$

$$\lambda P(z) + \mu[P(z) - p_0] = \frac{\lambda}{2} z^2 P(z) + \frac{\lambda}{2} z P(z) + \frac{\mu}{z}[P(z) - p_0]$$

$$P(z) = \frac{\mu p_0 \left(1 - \dfrac{1}{z}\right)}{\mu \left(1 - \dfrac{1}{z}\right) + \lambda \left(1 - \dfrac{z}{2} - \dfrac{z^2}{2}\right)}$$

$$P(z) = \frac{2\mu p_0}{2\mu - \lambda z(z + 2)}$$

Note that $\overline{\lambda} = \frac{3}{2}\lambda$ and so $\rho = \overline{\lambda}\overline{x} = \frac{3}{2}(\lambda/\mu)$. Thus

$$P(z) = \frac{p_0}{1 - \frac{1}{3}\rho z(z + 2)}$$

Also $P(1) = 1 = p_0/(1 - \rho)$ implies that $p_0 = 1 - \rho$. Therefore

$$P(z) = \frac{1 - \rho}{1 - \frac{1}{3}\rho z(z + 2)}$$

(d) Noting that

$$\frac{dP(z)}{dz} = (1 - \rho)\frac{\frac{1}{3}\rho(2z + 2)}{[1 - \frac{1}{3}\rho z(z + 2)]^2}$$

we have

$$\overline{N} = \frac{dP(z)}{dz}\bigg|_{z=1} = \frac{4}{3}\frac{\rho}{1 - \rho}$$

(e) Equation (1.82) says that

$$P(z) = \frac{\mu(1 - \rho)(1 - z)}{\mu(1 - z) - \lambda z[1 - G(z)]}$$

In our case, $g_1 = g_2 = \frac{1}{2}$ and therefore $G(z) = \frac{1}{2}z + \frac{1}{2}z^2$. Thus $G^{(1)}(z) = \frac{1}{2} + z$ and $\rho = \lambda G^{(1)}(1)/\mu = \frac{3}{2}(\lambda/\mu)$. Substituting these values into Eq. (1.82) gives

$$P(z) = \frac{\mu(1 - \rho)(1 - z)}{\mu(1 - z) - \lambda z(1 - \frac{1}{2}z - \frac{1}{2}z^2)}$$

Simplifying we have

$$P(z) = \frac{1 - \rho}{1 - \frac{1}{3}\rho z(z + 2)}$$

as before.   □

PROBLEM 4.7

Consider a bulk arrival system as described for Eq. (1.82) for which

$$g_i = (1 - \alpha)\alpha^i \qquad i = 0, 1, 2, \dots$$

(we assume $0 < \alpha < 1$). Find p_k = equilibrium probability of finding k in the system.

SOLUTION

Since $g_i = (1 - \alpha)\alpha^i, i = 0, 1, 2, \dots$, the appropriate z-transform for the distribution of bulk size is

$$G(z) = \sum_{i=0}^{\infty} g_i z^i = \sum_{i=0}^{\infty} (1 - \alpha)\alpha^i z^i$$

$$G(z) = \frac{1 - \alpha}{1 - \alpha z}$$

Further, we have $\rho = \lambda G^{(1)}(1)/\mu = \lambda\alpha/\mu(1 - \alpha)$. Therefore Eq. (1.82) gives

$$P(z) = \frac{\mu(1 - \rho)(1 - z)}{\mu(1 - z) - \lambda z[1 - G(z)]}$$

$$= \frac{\mu(1 - \rho)(1 - z)}{\mu(1 - z) - \lambda z\left[1 - \dfrac{1 - \alpha}{1 - \alpha z}\right]}$$

$$= \frac{\mu(1 - \rho)(1 - \alpha z)}{\mu(1 - \alpha z) - \lambda\alpha z}$$

$$P(z) = (1 - \rho)\frac{1 - \alpha z}{1 - \dfrac{\mu + \lambda}{\mu}\alpha z}$$

To invert $P(z)$, we note that the numerator degree equals the denominator degree, and so we divide once to bring the expression into proper form as follows:

$$P(z) = (1 - \rho)\left[\frac{\mu}{\lambda + \mu} + \frac{\lambda}{\lambda + \mu}\frac{1}{1 - \dfrac{\mu + \lambda}{\mu}\alpha z}\right]$$

By inspection, we have the final answer

$$p_0 = 1 - \rho$$

$$p_k = (1 - \rho)\cdot\frac{\lambda}{\lambda + \mu}\left(\frac{\mu + \lambda}{\mu}\alpha\right)^k \qquad k \geq 1 \qquad\qquad \square$$

PROBLEM 4.8

For the bulk arrival system described in the paragraph leading up to Eq. (1.82), find the mean $\overline{N}$ and variance σ_N^2 for the number of customers in the system. Express your answers in terms of the moments of the bulk arrival distribution.

SOLUTION

Equation (1.82) gives

$$P(z) = \frac{\mu(1-\rho)(1-z)}{\mu(1-z) - \lambda z[1 - G(z)]} = \frac{N(z)}{D(z)}$$

Let the kth moment of the bulk size be denoted by $\overline{g^k}$. Note that $\overline{g} = G^{(1)}(1)$ and $\overline{g^2} - \overline{g} = G^{(2)}(1)$. Since $\overline{g}$ customers arrive, on average, at each arrival instant, and since such "instants" occur at a rate λ, we have $\rho = \lambda\overline{g}/\mu$. To find $\overline{N}$ we use

$$\overline{N} = \left.\frac{dP(z)}{dz}\right|_{z=1} = \left.\frac{D(z)N^{(1)}(z) - N(z)D^{(1)}(z)}{[D(z)]^2}\right|_{z=1}$$

This last is indeterminate [since $N(1) = D(1) = 0$], and so we must use L'Hospital's rule twice as follows:

$$\overline{N} = P^{(1)}(1) = \left.\frac{D(z)N^{(2)}(z) - N(z)D^{(2)}(z)}{2D^{(1)}(z)D(z)}\right|_{z=1}$$

$$= \left.\frac{D(z)N^{(3)}(z) + D^{(1)}(z)N^{(2)}(z) - N^{(1)}(z)D^{(2)}(z) - N(z)D^{(3)}(z)}{2D^{(2)}(z)D(z) + 2[D^{(1)}(z)]^2}\right|_{z=1}$$

Now since

$$N(z) = \mu(1-\rho)(1-z) \qquad\qquad N(1) = 0$$
$$N^{(1)}(z) = -\mu(1-\rho) \qquad\qquad N^{(1)}(1) = -\mu(1-\rho)$$
$$N^{(2)}(z) = 0 \qquad\qquad N^{(2)}(1) = 0$$
$$N^{(3)}(z) = 0 \qquad\qquad N^{(3)}(1) = 0$$

and also

$$D(z) = \mu(1-z) - \lambda z[1 - G(z)] \qquad\qquad D(1) = 0$$
$$D^{(1)}(z) = -\mu - \lambda + \lambda G(z) + \lambda z G^{(1)}(z) \qquad\qquad D^{(1)}(1) = -\mu(1-\rho)$$
$$D^{(2)}(z) = 2\lambda G^{(1)}(z) + \lambda z G^{(2)}(z) \qquad\qquad D^{(2)}(1) = \lambda(\overline{g^2} + \overline{g})$$
$$D^{(3)}(z) = 3\lambda G^{(2)}(z) + \lambda z G^{(3)}(z) \qquad\qquad D^{(3)}(1) = \lambda(\overline{g^3} - \overline{g})$$

we may substitute in our expression for $\overline{N}$ to obtain

$$\overline{N} = \frac{\mu(1 - \rho)\lambda(\overline{g^2} - \overline{g})}{2[-\mu(1 - \rho)]^2} = \frac{\lambda(\overline{g^2} - \overline{g})}{2\mu(1 - \rho)}$$

$$\overline{N} = \frac{\rho}{1 - \rho}\left[\frac{\overline{g^2} + \overline{g}}{2\overline{g}}\right]$$

For the variance, we use

$$\sigma_N^2 = \left.\frac{d^2P(z)}{dz^2}\right|_{z=1} + \overline{N} - (\overline{N})^2$$

We see that

$$P^{(2)}(1) = \left.\frac{D(z)[D(z)N^{(2)}(z) - N(z)D^{(2)}(z)]}{[D(z)]^3}\right|_{z=1}$$

$$- \left.\frac{2D^{(1)}(z)[D(z)N^{(1)}(z) - N(z)D^{(1)}(z)]}{[D(z)]^3}\right|_{z=1}$$

We must now apply L'Hospital's rule three times, evaluate at $z = 1$, and eliminate the terms that vanish. We obtain

$$P^{(2)}(1) = \frac{-N^{(1)}(1)[2D^{(1)}(1)D^{(3)}(1) - 3[D^{(2)}(1)]^2]}{6[D^{(1)}(1)]^3}$$

$$= \frac{\mu(1 - \rho)[-2\mu(1 - \rho)\lambda(\overline{g^3} - \overline{g}) - 3\lambda^2(\overline{g^2} + \overline{g})^2]}{-6[\mu(1 - \rho)]^3}$$

$$= \frac{\lambda(\overline{g^3} - \overline{g})}{3\mu(1 - \rho)} + \frac{\lambda^2(\overline{g^2} + \overline{g})^2}{2\mu^2(1 - \rho)^2}$$

Hence

$$\sigma_N^2 = \frac{\lambda(\overline{g^3} - \overline{g})}{3\mu(1 - \rho)} + \frac{\lambda^2(\overline{g^2} + \overline{g})^2}{2\mu^2(1 - \rho)^2} + \frac{\lambda(\overline{g^2} + \overline{g})}{2\mu(1 - \rho)} - \frac{\lambda^2(\overline{g^2} + \overline{g})^2}{4\mu^2(1 - \rho)^2}$$

Thus

$$\sigma_N^2 = \frac{\rho}{1 - \rho}\left(\frac{2\overline{g^3} + 3\overline{g^2} + \overline{g}}{6\overline{g}}\right) + \frac{\rho^2}{(1 - \rho)^2}\left(\frac{\overline{g^2} + \overline{g}}{2\overline{g}}\right)^2$$

or

$$\sigma_N^2 = \frac{\rho}{1 - \rho}\cdot\left(\frac{2\overline{g^3} + 3\overline{g^2} + \overline{g}}{6\overline{g}}\right) + (\overline{N})^2 \qquad \square$$

PROBLEM 4.9

Consider an M/M/1 system with the following variation. Whenever the server becomes free, he accepts *two* customers (if at least two are available) from the queue into
service simultaneously. Of these two customers, only one receives service; when the
service for this one is completed, both customers depart (and so the other customer
got a "free ride").

If only one customer is available in the queue when the server becomes free, then
that customer is accepted alone and is serviced; if a new customer happens to arrive
when this single customer is being served, then the new customer joins the old one
in service and this new customer receives a "free ride."

In all cases, the service time is exponentially distributed with mean $1/\mu$ sec and
the average (Poisson) arrival rate is λ customers per second.

  (a) Draw the appropriate state diagram.
  (b) Write down the appropriate difference equations for p_k = equilibrium prob-
       ability of finding k customers in the system.
  (c) Solve for $P(z)$ in terms of p_0 and p_1.
  (d) Express p_1 in terms of p_0.
  (e) Find p_k for $k \geq 0$.

SOLUTION

  (a) The state-transition-rate diagram is

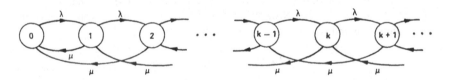

  (b) The equilibrium equations are

$$\lambda p_0 = \mu p_1 + \mu p_2 \qquad k = 0$$

$$(\lambda + \mu)p_k = \lambda p_{k-1} + \mu p_{k+2} \qquad k \geq 1$$

  (c) Multiply the kth equation by z^k and sum:

$$\lambda \sum_{k=0}^{\infty} p_k z^k + \mu \sum_{k=1}^{\infty} p_k z^k = \lambda \sum_{k=1}^{\infty} p_{k-1} z^k + \mu \sum_{k=0}^{\infty} p_{k+2} z^k + \mu p_1$$

$$\lambda P(z) + \mu[P(z) - p_0] = \lambda z P(z) + \frac{\mu}{z^2}[P(z) - p_0 - p_1 z] + \mu p_1$$

$$P(z) = \frac{\mu p_0(z + 1) + \mu p_1 z}{\mu(z + 1) - \lambda z^2}$$

(d) Since $P(1) = 1 = (2\mu p_0 + \mu p_1)/(2\mu - \lambda)$ we have

$$p_1 = 2(1 - p_0) - \frac{\lambda}{\mu}$$

Thus

$$P(z) = \frac{(2\mu - \lambda)z + \mu p_0(1 - z)}{\mu(z + 1) - \lambda z^2}$$

(e) We now solve for p_k explicitly. The two roots of the denominator of $P(z)$ are

$$z_1 = \frac{\mu - \sqrt{\mu^2 + 4\lambda\mu}}{2\lambda}, \quad z_2 = \frac{\mu + \sqrt{\mu^2 + 4\lambda\mu}}{2\lambda}$$

For $\lambda > 0$, we have

$$|z_1| = -z_1 = \frac{\sqrt{\mu^2 + 4\lambda\mu} - \mu}{2\lambda} < \frac{\sqrt{\mu^2 + 4\lambda\mu + 4\lambda^2} - \mu}{2\lambda} = 1$$

Thus the analyticity of $P(z)$ for $|z| \le 1$ requires that the one root of the numerator of $P(z)$ must be the denominator root z_1 (as $|z_1| < 1$), and so an equilibrium solution exists if and only if the other denominator root z_2 satisfies $|z_2| > 1$. We see that $|z_2| > 1$ if $\lambda < 2\mu$ as follows:

$$|z_2| = z_2 = \frac{\mu + \sqrt{\mu^2 + 4\lambda\mu}}{2\lambda} > \frac{(\lambda/2) + \sqrt{(\lambda^2/4) + 2\lambda^2}}{2\lambda} = 1$$

On the other hand, if $\lambda \ge 2\mu$, then $|z_2| \le 1$; hence we observe that the system is stable iff $\lambda < 2\mu$. Now let us find p_k. For $\lambda < 2\mu$, z_1 is the root of the numerator and so

$$(2\mu - \lambda)z_1 + \mu p_0(1 - z_1) = 0$$

or

$$p_0 = \frac{(2\mu - \lambda)z_1}{\mu(z_1 - 1)}$$

Substituting this into $P(z)$ (and canceling the root at $z = z_1$) we obtain

$$P(z) = \frac{2\mu - \lambda}{\lambda(1 - z_1)z_2} \frac{1}{1 - (z/z_2)}$$

But the product of the roots is simply $z_1 z_2 = -\mu/\lambda$; using this and the fact that $|z_2| > 1$ we invert by inspection to obtain

$$p_k = \frac{2\mu - \lambda}{\mu + \lambda z_2} \left(\frac{1}{z_2}\right)^k \qquad k \ge 0$$

as the desired solution.   □

PROBLEM 4.10

We consider the polynomial in Eq. (1.84) and wish to establish that exactly $r - 1$ roots lie in the range $|z| < 1$, and one root, say, z_0, lies in the region $|z_0| > 1$.

(a) Of the $r + 1$ roots, one occurs at $z = 1$. Use Rouché's theorem to show that exactly r roots lie in the unit disk $|z| \leq 1$.

(b) Show that $z = 1$ is the only root on the unit circle $|z| = 1$.

SOLUTION

(a) Equation (1.84) is

$$r\rho z^{r+1} - (1 + r\rho)z^r + 1 = 0$$

We split this function into two parts, $f(z) + g(z)$, where

$$f(z) = -(1 + r\rho)z^r, \quad g(z) = r\rho z^{r+1} + 1$$

For $\rho < 1$, we choose any real δ such that $0 < \delta < (1 - \rho)/r\rho$. (Note that $0 < 1 - \rho - r\rho\delta$.) Define the closed contour C as a circle about the origin of radius $1 + \delta$. On C, $|z| = 1 + \delta$, so we have

$$|f(z)| = |-(1 + r\rho)z^r| = (1 + r\rho)(1 + \delta)^r$$

and

$$|g(z)| = |r\rho z^{r+1} + 1| \leq r\rho(1 + \delta)^{r+1} + 1$$

Thus, on the contour C, we have

$$|f(z)| - |g(z)| \geq (1 + r\rho)(1 + \delta)^r - r\rho(1 + \delta)^{r+1} - 1$$

$$= (1 + \delta)^r(1 - r\rho\delta) - 1$$

$$\geq (1 + r\delta)(1 - r\rho\delta) - 1 \quad \text{[since } (1 + \delta)^r \geq 1 + r\delta]$$

$$= r\delta(1 - \rho - r\rho\delta)$$

$$> 0$$

So $|f(z)| > |g(z)|$ on C, and by Rouché's theorem, the denominator polynomial $f(z) + g(z)$ has exactly r roots in the range $|z| < 1 + \delta$ since $f(z)$ has. As this is true for all $0 < \delta < (1 - \rho)/r\rho$, letting $\delta \to 0$, we see that the denominator polynomial has exactly r roots in the range $|z| \leq 1$.

(b) Assume $|z| = 1$ and $r\rho z^{r+1} - (1 + r\rho)z^r + 1 = 0$. So $(1 + r\rho - r\rho z)z^r = 1$ and thus $|1 + r\rho(1 - z)| = 1$. Define $h = 1 + r\rho(1 - z)$ and so $|h| = 1$. Also

$-1 \leq \mathrm{Re}(z) \leq 1$, and so

$$\mathrm{Re}(h) = 1 + r\rho[1 - \mathrm{Re}(z)] \geq 1$$

Since $|h| = 1$ and $\mathrm{Re}(h) \geq 1$, we must have $h = 1$. Therefore

$$1 = h = 1 + r\rho(1 - z)$$

and we conclude that $z = 1$.   □

PROBLEM 4.11

Show that the solution to Eq. (1.97) gives a set of variables $\{x_i\}$ that guarantee that Eq. (1.96) is indeed the solution to Eq. (1.93).

SOLUTION

We proceed by showing that for Eq. (1.93) the left-hand side (LHS) equals the right-hand side (RHS) under the proposed solution Eq. (1.96) if the $\{x_i\}$ satisfy Eq. (1.97). Noting that $\delta_{k_i-1}\alpha_i(k_i) = \alpha_i(k_i)$ for each i, the LHS can be written as

$$\mathrm{LHS} = p(k_1, \ldots, k_N) \sum_{i=1}^{N} \alpha_i(k_i)\mu_i$$

The RHS is

$$\sum_{i=1}^{N} \sum_{j=1}^{N} \delta_{k_j-1}\alpha_i(k_i + 1)\mu_i r_{ij} p(k_1, \ldots, k_j - 1, \ldots, k_i + 1, \ldots, k_N)$$

Using the proposed solution [Eq. (1.96)]

$$p(k_1, \ldots, k_N) = \frac{1}{G(K)} \prod_{l=1}^{N} \frac{x_l^{k_l}}{\beta_l(k_l)}$$

we may write

$$p(k_1, \ldots, k_j - 1, \ldots, k_i + 1, \ldots, k_N) = \frac{1}{G(K)} \prod_{l=1}^{N} \frac{x_l^{k_l}}{\beta_l(k_l)} \cdot \frac{\dfrac{x_i}{x_j}}{\dfrac{\beta_i(k_i + 1)}{\beta_i(k_i)} \cdot \dfrac{\beta_j(k_j - 1)}{\beta_j(k_j)}}$$

$$= p(k_1, \ldots, k_N) \frac{x_i}{x_j} \cdot \frac{\beta_i(k_i)}{\beta_i(k_i + 1)} \cdot \frac{\beta_j(k_j)}{\beta_j(k_j - 1)}$$

Since $\beta_i(k_i + 1) = \beta_i(k_i)\alpha_i(k_i + 1)$ and $\beta_j(k_j) = \beta_j(k_j - 1)\alpha_j(k_j)$ we have

$$p(k_1, \ldots, k_j - 1, \ldots, k_i + 1, \ldots, k_N) = p(k_1, \ldots, k_N) \frac{x_i}{x_j} \cdot \frac{\alpha_j(k_j)}{\alpha_i(k_i + 1)}$$

Thus the RHS may be expressed as

$$\text{RHS} = \sum_{i=1}^{N}\sum_{j=1}^{N}\delta_{k_j-1}\alpha_i(k_i+1)\mu_i r_{ij}p(k_1,\dots,k_N)\frac{x_i}{x_j}\cdot\frac{\alpha_j(k_j)}{\alpha_i(k_i+1)}$$

$$= p(k_1,\dots,k_N)\sum_{i=1}^{N}\sum_{j=1}^{N}\delta_{k_j-1}\alpha_j(k_j)\mu_i r_{ij}\frac{x_i}{x_j}$$

But since $\delta_{k_j-1}\alpha_j(k_j)=\alpha_j(k_j)$,

$$\text{RHS} = p(k_1,\dots,k_N)\sum_{i=1}^{N}\sum_{j=1}^{N}\alpha_j(k_j)\mu_i r_{ij}\frac{x_i}{x_j}$$

Thus the LHS equals the RHS if

$$\sum_{i=1}^{N}\alpha_i(k_i)\mu_i = \sum_{i=1}^{N}\sum_{j=1}^{N}\alpha_j(k_j)\mu_i r_{ij}\frac{x_i}{x_j}$$

We prove this last equality by applying Eq. (1.97). That is,

$$\mu_i = \sum_{j=1}^{N}\mu_j r_{ji}\frac{x_j}{x_i}\qquad i=1,2,\dots,N$$

Multiplying each of these N equations by $\alpha_i(k_i)$, summing on i, and then interchanging the indices i and j gives

$$\sum_{i=1}^{N}\alpha_i(k_i)\mu_i = \sum_{i=1}^{N}\sum_{j=1}^{N}\alpha_j(k_j)\mu_i r_{ij}\frac{x_i}{x_j}$$

which was to be shown. This completes the proof.   □

PROBLEM 4.12

Consider a two-node Markovian queueing network (of the more general type considered by Jackson) for which $N=2$, $m_1=m_2=1$, $\mu_{k_i}=\mu_i$ (constant service rate), and which has transition probabilities (r_{ij}) as described in the following matrix:

$r_{ij}=$ i	j 0	1	2	3
0	0	1	0	0
1	0	0	$1-\alpha$	α
2	0	1	0	0

where $0 < \alpha < 1$ and nodes 0 and $N + 1$ are the "source" and "sink" nodes, respectively. We also have (for some integer K)

$$\gamma(k_1 + k_2) = \begin{cases} \infty & k_1 + k_2 \neq K \\ 0 & k_1 + k_2 = K \end{cases}$$

and assume the system initially contains K customers.

(a) Find e_i $(i = 1, 2)$ as given in Eq. (1.92).

(b) Since $N = 2$, let us denote $p(k_1, k_2) = p(k_1, K - k_1)$ by p_{k_1}. Find the balance equations for p_{k_1}.

(c) Solve these equations for p_{k_1} explicitly.

(d) By considering the fraction of time the first node is busy, find the time between customer departures from the network (via node 1, of course).

SOLUTION

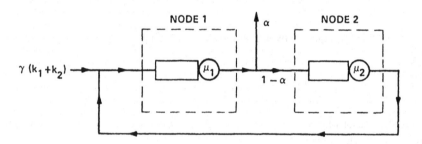

(a) Using Eq. (1.92) e_i is found as follows. From

$$e_1 = r_{01} + \sum_{j=1}^{2} e_j r_{j1}$$

we have

$$e_1 = 1 + e_1 \cdot 0 + e_2 \cdot 1 = 1 + e_2$$

Also, from

$$e_2 = r_{02} + \sum_{j=1}^{2} e_j r_{j2}$$

we have

$$e_2 = 0 + e_1 \cdot (1 - \alpha) + e_2 \cdot 0 = e_1(1 - \alpha)$$

Solving gives

$$e_1 = \frac{1}{\alpha}, \; e_2 = \frac{1 - \alpha}{\alpha}$$

(b) Since $\gamma(k_1 + k_2) = 0$ for $k_1 + k_2 = K$, no one enters the system if K customers are already there. But as soon as a departure takes place, another customer immediately enters the system [since $\gamma(k_1 + k_2) = \infty$ for $k_1 + k_2 \neq K$]. Thus we have a *closed* queueing network as follows:

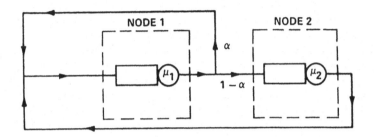

For this network we have the following state diagram (labeling the states only by the number k_1 present at node 1):

The balance equations for p_{k_1} are

$$\mu_2 p_0 = (1 - \alpha)\mu_1 p_1 \qquad k_1 = 0$$

$$[\mu_2 + (1 - \alpha)\mu_1]p_{k_1} = \mu_2 p_{k_1 - 1} + (1 - \alpha)\mu_1 p_{k_1 + 1} \qquad 0 < k_1 < K$$

$$(1 - \alpha)\mu_1 p_K = \mu_2 p_{K - 1} \qquad k_1 = K$$

(c) Noting that this system has the same state diagram as does M/M/1/K (see also Problem 3.12), we immediately solve for p_{k_1}:

$$p_{k_1} = \frac{1 - \dfrac{\mu_2}{(1 - \alpha)\mu_1}}{1 - \left[\dfrac{\mu_2}{(1 - \alpha)\mu_1}\right]^{K+1}} \left[\dfrac{\mu_2}{(1 - \alpha)\mu_1}\right]^{k_1} \qquad 0 \le k_1 \le K$$

(d) The first node is busy a fraction $(1 - p_0)$ of the time. For a very long time interval τ, the first node is busy for $(1 - p_0)\tau$ seconds. While node 1 is busy, customers leave the system at rate $\alpha\mu_1$. Thus $\alpha\mu_1(1 - p_0)\tau$ is the average number of departures during τ. Therefore, by renewal theory arguments, the average time between departures will be $1/\alpha\mu_1(1 - p_0)$ or, upon substituting for p_0, we find

$$E[\text{interdeparture time}] = \frac{1 - \alpha}{\alpha\mu_2} \frac{1 - \left[\dfrac{\mu_2}{(1 - \alpha)\mu_1}\right]^{K+1}}{1 - \left[\dfrac{\mu_2}{(1 - \alpha)\mu_1}\right]^{K}}. \qquad \square$$

CHAPTER 5

THE QUEUE M/G/1

PROBLEM 5.1

Given that

$$\hat{F}^*(s) = \frac{1 - F^*(s)}{sm_1}$$

as discussed in the footnote on page 22, prove that the kth moment of the residual life, denoted by r_k, is given by

$$r_k = \frac{m_{k+1}}{(k + 1)m_1}$$

where m_k is the kth moment of the interval length as discussed in that footnote.

SOLUTION

We first derive a power series expansion for the Laplace transform of any non-negative random variable in terms of its moments. To this end, let $X \geq 0$ be a random variable with density $h(x)$, Laplace transform $H^*(s)$, and kth moment $\overline{X^k}$. Expanding e^{-sx} in a power series, the Laplace transform is given by

$$H^*(s) \overset{\Delta}{=} \int_0^\infty e^{-sx} h(x)\, dx = \int_0^\infty \sum_{k=0}^\infty \frac{(-sx)^k}{k!} h(x)\, dx$$

$$= \sum_{k=0}^\infty \frac{(-1)^k s^k}{k!} \int_0^\infty x^k h(x)\, dx = \sum_{k=0}^\infty \frac{(-1)^k \overline{X^k}}{k!} s^k$$

or

$$H^*(s) = 1 + \sum_{k=1}^{\infty} \frac{(-1)^k \overline{X^k}}{k!} s^k$$

[Note that this immediately gives $H^{*(n)}(0) = (-1)^n \overline{X^n}$.] Applying the above expansion to $\hat{F}^*(s)$, the transform of the residual life density, we find

$$\hat{F}^*(s) = 1 + \sum_{k=1}^{\infty} \frac{(-1)^k r_k}{k!} s^k$$

But we know that

$$\hat{F}^*(s) = \frac{1 - F^*(s)}{s m_1}$$

Using the expansion for $F^*(s)$ in this equation, we have

$$\hat{F}^*(s) = \frac{1 - \left[1 + \sum_{k=1}^{\infty} \frac{(-1)^k m_k}{k!} s^k\right]}{s m_1} = \sum_{k=0}^{\infty} \frac{(-1)^k m_{k+1}}{(k+1)! m_1} s^k$$

or

$$\hat{F}^*(s) = 1 + \sum_{k=1}^{\infty} \frac{(-1)^k m_{k+1}}{(k+1)! m_1} s^k$$

Equating coefficients of these two power series expansions for $\hat{F}^*(s)$ gives

$$r_k = \frac{m_{k+1}}{(k+1) m_1} \qquad k = 1, 2, 3, \ldots \qquad \square$$

PROBLEM 5.2

Here we derive the residual lifetime density $\hat{f}(x)$ discussed in the footnote on page 22. We use the notation from that footnote. In particular, recall that t is the randomly selected point on the time axis on which the sequence of instants τ_k has been defined, and that Y is the residual life of the interval in which t falls.

(a) Observing that the event $\{Y \leq y\}$ can occur if and only if $t < \tau_k \leq t + y < \tau_{k+1}$ for some k, show that

$$\hat{F}_t(y) \triangleq P[Y \leq y]$$

$$= \sum_{k=1}^{\infty} \int_t^{t+y} [1 - F(t + y - x)] \, dP[\tau_k \leq x]$$

(b) Observing that $\tau_k \leq x$ if and only if $\alpha(x)$, the number of "arrivals" in $(0, x)$, is at least k, that is, $P[\tau_k \leq x] = P[\alpha(x) \geq k]$, show that

$$\sum_{k=1}^{\infty} P[\tau_k \leq x] = \sum_{k=1}^{\infty} kP[\alpha(x) = k]$$

(c) Let $\hat{F}(y) = \lim \hat{F}_t(y)$ as $t \to \infty$ with corresponding pdf $\hat{f}(y)$. Show that we now have

$$\hat{f}(y) = \frac{1 - F(y)}{m_1}$$

[HINT: Use the key renewal theorem (see [FELL 66] or [TAKA 62]).]

SOLUTION

(a) The event $\{Y \leq y\}$ may be written as the following disjoint union of events

$$\{Y \leq y\} = \bigcup_{k=1}^{\infty} \{t < \tau_k \leq t + y < \tau_{k+1}\}$$

That is, the instant $t + y$ must be separated from t by *at least* one arrival. We let τ_k denote the latest of these arrival instants (therefore these instants must satisfy $t < \cdots < \tau_{k-1} < \tau_k \leq t + y < \tau_{k+1}$).

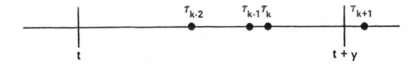

Rewriting this union we see that

$$\{Y \leq y\} = \bigcup_{k=1}^{\infty} \{t < \tau_k \leq t + y, \ t + y - \tau_k < \tau_{k+1} - \tau_k\}$$

$$= \bigcup_{k=1}^{\infty} \{t < \tau_k \leq t + y, \ t + y - \tau_k < t_{k+1}\}$$

where we use the definition $t_{k+1} \overset{\Delta}{=} \tau_{k+1} - \tau_k$, which is the length of the

$(k + 1)$th interarrival interval. Thus

$$\hat{F}_t(y) = P[Y \le y] = \sum_{k=1}^{\infty} P[t < \tau_k \le t + y, \ t_{k+1} > t + y - \tau_k]$$

Conditioning on τ_k we obtain

$$\hat{F}_t(y) = \sum_{k=1}^{\infty} \int_0^{\infty} P[t < \tau_k \le t + y, \ t_{k+1} > t + y - \tau_k \mid \tau_k = x]$$

$$\times P[x < \tau_k \le x + dx]$$

$$= \sum_{k=1}^{\infty} \int_0^{\infty} P[t < x \le t + y, \ t_{k+1} > t + y - x] \, dP[\tau_k \le x]$$

$$= \sum_{k=1}^{\infty} \int_t^{t+y} P[t_{k+1} > t + y - x] \, dP[\tau_k \le x]$$

Thus

$$\hat{F}_t(y) = \sum_{k=1}^{\infty} \int_t^{t+y} [1 - F(t + y - x)] \, dP[\tau_k \le x]$$

[Note: A similar formula for $\hat{F}_t(y)$ may be obtained as follows. Recognizing that the event $\{Y \le y\}$ may be written as

$$\{Y \le y\} = \bigcup_{k=0}^{\infty} \{\tau_k < t, \ t \le \tau_{k+1} < t + y\}$$

(where $\tau_0 \overset{\Delta}{=} 0$), we then have

$$\hat{F}_t(y) = \sum_{k=0}^{\infty} P[\tau_k < t, \ t - \tau_k \le t_{k+1} < t + y - \tau_k]$$

$$= \sum_{k=0}^{\infty} \int_0^{\infty} P[x < t, \ t - x \le t_{k+1} < t + y - x] \, dP[\tau_k \le x]$$

$$= \sum_{k=0}^{\infty} \int_0^{t} P[t - x \le t_{k+1} < t + y - x] \, dP[\tau_k \le x]$$

or

$$\hat{F}_t(y) = F(t + y) - F(t) + \sum_{k=1}^{\infty} \int_0^{t} [F(t + y - x) - F(t - x)] \, dP[\tau_k \le x]$$

It can be shown that this expression for $\hat{F}_t(y)$ is equal to that derived immediately above.]

(b) Recall that for a discrete random variable $X \geq 0$, we have

$$E[X] = \sum_{j=1}^{\infty} jP[X = j] = \sum_{j=1}^{\infty} \sum_{k=1}^{j} P[X = j]$$

$$= \sum_{k=1}^{\infty} \sum_{j=k}^{\infty} P[X = j] = \sum_{k=1}^{\infty} P[X \geq k]$$

We are given $P[\tau_k \leq x] = P[\alpha(x) \geq k]$. Summing on k and using our result above for $E[X]$, we have

$$\sum_{k=1}^{\infty} P[\tau_k \leq x] = \sum_{k=1}^{\infty} P[\alpha(x) \geq k]$$

$$= \sum_{k=1}^{\infty} kP[\alpha(x) = k] = E[\alpha(x)]$$

(c) We can note, using the Strong Law of Large Numbers (see [KLEI 75]), that as $x \to \infty$ then $\alpha(x)/x \to 1/m_1$ with probability 1. We also have $E[\alpha(x)]/x \to 1/m_1$ as $x \to \infty$, which is simply the Elementary Renewal Theorem. This gives the intuitive result

$$\sum_{k=1}^{\infty} P[\tau_k \leq x] = E[\alpha(x)] \approx \frac{x}{m_1} \qquad \text{for large } x$$

To give a rigorous proof of part (c), we proceed as follows. We first simplify $\hat{F}_t(y)$ as given in part (a):

$$\hat{F}_t(y) = \int_t^{t+y} [1 - F(t + y - x)] d \sum_{k=1}^{\infty} P[\tau_k \leq x]$$

$$= \int_t^{t+y} [1 - F(t + y - x)] dE[\alpha(x)] \qquad \text{[from part (b)]}$$

$$= \int_0^{t+y} [1 - F(t + y - x)] dE[\alpha(x)]$$

$$- \int_0^{t} [1 - F(t + y - x)] dE[\alpha(x)]$$

We next wish to show that for any $0 \leq u < \infty$,

$$\int_0^{u} [1 - F(u - x)] dE[\alpha(x)] = F(u)$$

To this end we note that

$$E[\alpha(x)] = \sum_{k=1}^{\infty} P[\tau_k \leq x] = \sum_{k=1}^{\infty} F_{(k)}(x)$$

where $F_{(1)}(x) = F(x)$ and, for $k \geq 1$, $F_{(k+1)}(x)$ is the PDF of τ_{k+1} $(= \tau_k + t_{k+1})$ defined by $F_{(k+1)}(x) = \int_0^x F(x - y) \, dF_{(k)}(y)$. Thus

$$
\int_0^u [1 - F(u - x)] \, dE[\alpha(x)] = \int_0^u [1 - F(u - x)] \, d \sum_{k=1}^{\infty} F_{(k)}(x)
$$

$$
= \sum_{k=1}^{\infty} \int_0^u [1 - F(u - x)] \, dF_{(k)}(x)
$$

$$
= \sum_{k=1}^{\infty} \left[F_{(k)}(u) - F_{(k+1)}(u) \right]
$$

$$
= \lim_{k \to \infty} \left[F_{(1)}(u) - F_{(k+1)}(u) \right] = F(u)
$$

as desired. Replacing u by $t + y$ in our earlier equation gives

$$
F(t + y) = \int_0^{t+y} [1 - F(t + y - x)] \, dE[\alpha(x)]
$$

We now apply this result to $\hat{F}_t(y)$ as expressed as the difference of integrals above and obtain

$$
\hat{F}_t(y) = F(t + y) - \int_0^t [1 - F(t + y - x)] \, dE[\alpha(x)]
$$

and so

$$
\hat{F}(y) \overset{\Delta}{=} \lim_{t \to \infty} \hat{F}_t(y) = 1 - \lim_{t \to \infty} \int_0^t [1 - F(t + y - x)] \, dE[\alpha(x)]
$$

To evaluate this limit we use the key renewal theorem of Smith, which may be found in [FELL 66] pages 347–351 (see also [TAKA 62] page 227). The key renewal theorem states:

If F is not lattice, and if $h(t)$ is directly Riemann integrable, then

$$
\lim_{t \to \infty} \int_0^t h(t - x) \, dE[\alpha(x)] = \frac{1}{m_1} \int_0^{\infty} h(t) \, dt
$$

We apply this theorem to the function $h(t) = 1 - F(t + y)$ to obtain

$$
\lim_{t \to \infty} \int_0^t [1 - F(t + y - x)] \, dE[\alpha(x)] = \frac{1}{m_1} \int_0^{\infty} [1 - F(t + y)] \, dt
$$

Then

$$\hat{F}(y) = 1 - \frac{1}{m_1} \int_0^\infty [1 - F(t + y)] \, dt$$

$$= \frac{1}{m_1} \left[\int_0^\infty [1 - F(t)] \, dt - \int_y^x [1 - F(t)] \, dt \right]$$

$$\hat{F}(y) = \frac{1}{m_1} \int_0^y [1 - F(t)] \, dt$$

and so, finally,

$$\hat{f}(y) = \frac{1 - F(y)}{m_1} \qquad\qquad \square$$

PROBLEM 5.3

Let us rederive the P-K mean-value formula (1.101).

(a) Recognizing that a new arrival is delayed by one service time for each queued customer plus the residual service time of the customer in service, write an expression for W in terms of $\overline{N}_q$, ρ, $\overline{x}$, and $\overline{x^2}$.

(b) Use Little's result in (a) to obtain Eq. (1.101).

SOLUTION

(a) Since $r_k = p_k = d_k$ for M/G/1, the average number in queue that an arrival finds is $\overline{N}_q$, the average number in queue (over all time). Thus we may write

$$W = \overline{N}_q \overline{x} + \frac{\overline{x^2}}{2\overline{x}} P[\tilde{w} > 0]$$

But for Poisson arrivals, $P[\tilde{w} > 0] = P[\text{busy system}] = \rho$. Thus

$$W = \overline{N}_q \overline{x} + \frac{\overline{x^2}}{2\overline{x}} \rho$$

(b) By Little's result, $\overline{N}_q = \lambda W$. So

$$W = \rho W + \frac{\overline{x^2}}{2\overline{x}} \rho$$

Thus

$$W = \frac{\lambda \overline{x^2}/2}{1 - \rho}$$

126 CHAPTER 5 THE QUEUE M/G/1

which is Eq. (1.101). Note further that, if we define the squared coefficient of variation for the service time as $C_b^2 = \sigma_b^2/(\bar{x})^2$, then we may rewrite Eq. (1.101) as

$$\frac{W}{\bar{x}} = \frac{\rho(1 + C_b^2)}{2(1 - \rho)} \qquad \Box$$

PROBLEM 5.4

Replace $1 - \rho$ in Eq. (1.104) by an unknown constant and show that $Q(1) = B^*(0) = 1$ easily gives us the correct value of $1 - \rho$ for this constant.

SOLUTION

We have

$$Q(z) = B^*(\lambda - \lambda z)\frac{K(1 - z)}{B^*(\lambda - \lambda z) - z}$$

or

$$\frac{Q(z)}{B^*(\lambda - \lambda z)} = K\frac{z - 1}{z - B^*(\lambda - \lambda z)}$$

Since $Q(1) = B^*(0) = 1$, we may evaluate the constant K by using L'Hospital's rule on the right-hand expression. This gives

$$1 = K\frac{\frac{d}{dz}(z - 1)\Big|_{z=1}}{\frac{d}{dz}(z - B^*(\lambda - \lambda z))\Big|_{z=1}} = K\frac{1}{1 - \lambda[-B^{*(1)}(0)]}$$

So $K = 1 - \lambda\bar{x} = 1 - \rho$. $\qquad \Box$

PROBLEM 5.5

(a) From Eq. (1.104) form $Q^{(1)}(1)$ and show that it gives the expression found in Eq. (1.102).

(b) From Eq. (1.105), find the first two moments of the waiting time and compare with Eqs. (1.101) and (1.109).

SOLUTION

(a) Equation (1.104) is $Q(z) = B^*(\lambda - \lambda z)(1 - \rho)(1 - z)/[B^*(\lambda - \lambda z) - z]$. We first define a function

$$f(z) = \frac{B^*(\lambda - \lambda z) - z}{1 - z}$$

This will permit us to factor $(1 - z)$ from the denominator of $Q(z)$, thereby eliminating the indeterminacies at $z = 1$ and simplifying our calculations. Applying the expansion procedure from Problem 5.1 to $B^*(\lambda - \lambda z)$, we see that

$$B^*(\lambda - \lambda z) = 1 - \bar{x}(\lambda - \lambda z) + \frac{\overline{x^2}}{2!}(\lambda - \lambda z)^2 - \frac{\overline{x^3}}{3!}(\lambda - \lambda z)^3 + \cdots$$

and thus

$$f(z) = 1 - \lambda\bar{x} + \frac{\lambda^2\overline{x^2}}{2!}(1 - z) - \frac{\lambda^3\overline{x^3}}{3!}(1 - z)^2 + \cdots$$

So we have $f(1) = 1 - \rho$, and $f^{(1)}(1) = -\lambda^2\overline{x^2}/2$. Noting that

$$Q(z) = (1 - \rho)\frac{B^*(\lambda - \lambda z)}{f(z)}$$

and hence

$$Q^{(1)}(z) = (1 - \rho)\frac{f(z)B^{*(1)}(\lambda - \lambda z)(-\lambda) - B^*(\lambda - \lambda z)f^{(1)}(z)}{[f(z)]^2}$$

we obtain

$$\bar{q} = Q^{(1)}(1) = (1 - \rho)\left[\frac{(1 - \rho)\lambda\bar{x} + \dfrac{\lambda^2\overline{x^2}}{2}}{(1 - \rho)^2}\right] = \rho + \frac{\lambda^2\overline{x^2}}{2(1 - \rho)}$$

Noting that $\bar{q} = \bar{N}$ for M/G/1, we recognize this as Eq. (1.102).

(b) Equation (1.105) is $W^*(s) = s(1 - \rho)/[s - \lambda + \lambda B^*(s)]$. In a fashion similar to part (a), define a function

$$f(s) = \frac{s - \lambda + \lambda B^*(s)}{s} = 1 - \frac{\lambda}{s} + \frac{\lambda}{s}B^*(s)$$

This permits us to remove the indeterminacy at $s = 0$. Using the expansion method from Problem 5.1 for $B^*(s)$, we find

$$f(s) = 1 - \frac{\lambda}{s} + \frac{\lambda}{s}\left[1 - s\bar{x} + \frac{s^2\overline{x^2}}{2!} - \frac{s^3\overline{x^3}}{3!} + \frac{s^4\overline{x^4}}{4!} - \cdots\right]$$

$$= 1 - \lambda\bar{x} + \frac{\lambda\overline{x^2}}{2!}s - \frac{\lambda\overline{x^3}}{3!}s^2 + \frac{\lambda\overline{x^4}}{4!}s^3 - \cdots$$

Therefore

$$f^{(1)}(s) = \frac{\lambda\overline{x^2}}{2!} - \frac{2\lambda\overline{x^3}}{3!}s + \frac{3\lambda\overline{x^4}}{4!}s^2 - \cdots$$

and

$$f^{(2)}(s) = -\frac{2\lambda \overline{x^3}}{3!} + \frac{(3 \cdot 2)\lambda \overline{x^4}}{4!}s - \cdots$$

Thus we have

$$f(0) = 1 - \rho, \quad f^{(1)}(0) = \frac{\lambda \overline{x^2}}{2}, \quad f^{(2)}(0) = -\frac{\lambda \overline{x^3}}{3}$$

Recall that $W^*(s) = (1 - \rho)/f(s)$. Differentiating,

$$W^{*(1)}(s) = -(1 - \rho)\frac{f^{(1)}(s)}{[f(s)]^2}$$

and

$$W^{*(2)}(s) = -(1 - \rho)\frac{[f(s)]^2 f^{(2)}(s) - 2f(s)[f^{(1)}(s)]^2}{[f(s)]^4}$$

Hence

$$W = -W^{*(1)}(0) = \frac{\lambda \overline{x^2}}{2(1 - \rho)}$$

which is Eq. (1.101). Also

$$\overline{w^2} = W^{*(2)}(0) = -(1 - \rho)\frac{(1 - \rho)^2\left(-\dfrac{\lambda \overline{x^3}}{3}\right) - 2(1 - \rho)\left(\dfrac{\lambda \overline{x^2}}{2}\right)^2}{(1 - \rho)^4}$$

or

$$\overline{w^2} = 2(\overline{w})^2 + \frac{\lambda \overline{x^3}}{3(1 - \rho)}$$

But $\sigma_w^2 = \overline{w^2} - (\overline{w})^2$, and so

$$\sigma_w^2 = W^2 + \frac{\lambda \overline{x^3}}{3(1 - \rho)}$$

which is Eq. (1.109). □

PROBLEM 5.6

We wish to prove that the limiting probability r_k for the number of customers found by an arrival is equal to the limiting probability d_k for the number of customers left behind by a departure, in any queueing system in which the state changes by unit step values only (positive or negative). Beginning at $t = 0$, let x_n be those instants when $N(t)$ (the number in system) increases by one and y_n be those instants when $N(t)$ decreases by unity, $n = 1, 2, \ldots$. Let $N(x_n^-)$ be denoted by α_n and $N(y_n^+)$ by β_n. Let $N(0) = i$.

(a) Show that if $\beta_{n+i} \leq k$, then $\alpha_{n+k+1} \leq k$.

(b) Show that if $\alpha_{n+k+1} \leq k$, then $\beta_{n+i} \leq k$.

(c) Show that (a) and (b) must therefore give, for any k,

$$\lim_{n \to \infty} P[\beta_n \leq k] = \lim_{n \to \infty} P[\alpha_n \leq k]$$

which establishes that $r_k = d_k$.

SOLUTION

(a) Assume $\beta_{n+i} \leq k$ (for some $k \geq 0$, $n > 0$); that is, the $(n + i)$th departure leaves behind at most k customers. Let

$$a \stackrel{\Delta}{=} \text{number of arrivals by time } y_{n+i}^+$$

$$= \text{number of arrivals prior to the } (n + i)\text{th departure}$$

Then $a + i - (n + i) = \beta_{n+i} \leq k$. Hence $a \leq n + k$, and thus the $(n + k + 1)$th arrival occurs after the $(n + i)$th departure; that is,

$$y_{n+i}^+ \leq x_{n+k+1}^-$$

Since there were at least $n + i$ departures before the $(n + k + 1)$th arrival, we have

$$\alpha_{n+k+1} \leq n + k + i - (n + i) = k$$

(b) Assume $\alpha_{n+k+1} \leq k$ (for some $k \geq 0$, $n > 0$); that is, the $(n + k + 1)$th arrival finds at most k customers. Let

$$d \stackrel{\Delta}{=} \text{number of departures by time } x_{n+k+1}^-$$

$$= \text{number of departures prior to the } (n + k + 1)\text{th arrival}$$

Then $n + k + i - d = \alpha_{n+k+1} \leq k$. Hence $n + i \leq d$ and the $(n + i)$th departure must have occurred before the $(n + k + 1)$th arrival; that is,

$$y_{n+i}^+ \leq x_{n+k+1}^-$$

Since there were at most $n + k$ arrivals by the $(n + i)$th departure, we see that

$$\beta_{n+i} \leq n + k + i - (n + i) = k$$

(c) From parts (a) and (b) (for $k \geq 0$)

$$\beta_{n+i} \leq k \text{ iff } \alpha_{n+k+1} \leq k \qquad (n > 0)$$

So

$$P[\beta_{n+i} \leq k] = P[\alpha_{n+k+1} \leq k]$$

Letting $n \to \infty$ (assuming the equilibrium probabilities exist), we have for fixed $k \geq 0$

$$\lim_{n \to \infty} P[\beta_n \leq k] = \lim_{n \to \infty} P[\beta_{n+i} \leq k]$$
$$= \lim_{n \to \infty} P[\alpha_{n+k+1} \leq k] = \lim_{n \to \infty} P[\alpha_n \leq k]$$

or

$$d_k = r_k \qquad \text{for all } k \geq 0 \qquad\qquad \square$$

PROBLEM 5.7

In this problem, we develop an alternative technique, the method of supplementary variables, to solve the M/G/1 queue. As usual, let $P_k(t) = P[N(t) = k]$. Moreover, let $P_k(t, x_0)\Delta x_0 = P[N(t) = k, x_0 < X_0(t) \leq x_0 + \Delta x_0]$, where $X_0(t)$ is the service already received by the customer in service at time t.

(a) Show that

$$\frac{\partial P_0(t)}{\partial t} = -\lambda P_0(t) + \int_0^{\infty} P_1(t, x_0) r(x_0) \, dx_0$$

where

$$r(x_0) = \frac{b(x_0)}{1 - B(x_0)}$$

(b) Let $p_k = \lim_{t \to \infty} P_k(t)$ and $p_k(x_0) = \lim_{t \to \infty} P_k(t, x_0)$. From (a) we have the equilibrium result

$$\lambda p_0 = \int_0^{\infty} p_1(x_0) r(x_0) \, dx_0$$

Show the following equilibrium results [where $p_0(x_0) \overset{\Delta}{=} 0$]:

(i) $\dfrac{\partial p_k(x_0)}{\partial x_0} = -[\lambda + r(x_0)] p_k(x_0) + \lambda p_{k-1}(x_0)$ $k \geq 1$

(ii) $p_k(0) = \displaystyle\int_0^{\infty} p_{k+1}(x_0) r(x_0) \, dx_0$ $k > 1$

(iii) $p_1(0) = \displaystyle\int_0^{\infty} p_2(x_0) r(x_0) \, dx_0 + \lambda p_0$

(c) The four equations in (b) determine the equilibrium probabilities when combined with an appropriate normalization equation. In terms of p_0 and $p_k(x_0)$ ($k = 1, 2, \ldots$) give this normalization equation.

(d) Let $R(z, x_0) = \sum_{k=1}^{\infty} p_k(x_0)z^k$. Show that

$$\frac{\partial R(z, x_0)}{\partial x_0} = [\lambda z - \lambda - r(x_0)]R(z, x_0)$$

and

$$zR(z, 0) = \int_0^{\infty} r(x_0)R(z, x_0)\,dx_0 + \lambda z(z - 1)p_0$$

(e) Show that the solution for $R(z, x_0)$ from (d) must be

$$R(z, x_0) = R(z, 0)e^{-\lambda x_0(1-z) - \int_0^{x_0} r(y)\,dy}$$

$$R(z, 0) = \frac{\lambda z(z - 1)p_0}{z - B^*(\lambda - \lambda z)}$$

(f) Defining $R(z) \overset{\Delta}{=} \int_0^{\infty} R(z, x_0)\,dx_0$, show that

$$R(z) = R(z, 0)\frac{1 - B^*(\lambda - \lambda z)}{\lambda(1 - z)}$$

(g) From the normalization equation of (c), now show that

$$p_0 = 1 - \rho \qquad (\rho = \lambda\bar{x})$$

(h) Since $R(z) = \sum_{k=1}^{\infty} p_k z^k$, we note that

$$Q(z) = p_0 + R(z)$$

Show that $Q(z)$ expressed this way is identical to the P-K transform equation (1.104). (See [COX 55] for additional details of this method.)

SOLUTION

(a) If we are to find zero customers at time $t + \Delta t$, then to within $o(\Delta t)$, it must be that at time t either (i) we had zero customers and there were no arrivals in $(t, t + \Delta t)$ or (ii) we had one customer at time t and this customer completed service in $(t, t + \Delta t)$. We calculate this latter case by conditioning on the service received, $X_0(t)$, and noting that the departure rate when $X_0(t) = x_0$ is simply $r(x_0) = b(x_0)/[1 - B(x_0)]$. Thus

$$P_0(t + \Delta t) = (1 - \lambda \Delta t)P_0(t) + \int_0^{\infty} P_1(t, x_0)\,dx_0\, r(x_0)\,\Delta t + o(\Delta t)$$

Subtracting $P_0(t)$ from both sides, dividing by Δt, and letting $\Delta t \to 0$ gives

$$\frac{\partial P_0(t)}{\partial t} = -\lambda P_0(t) + \int_0^{\infty} P_1(t, x_0)r(x_0)\,dx_0$$

(b) We first note that as $t \to \infty$ then $\partial P_0(t)/\partial t \to 0$, and thus the expression in part (a) becomes

$$\lambda p_0 = \int_0^\infty p_1(x_0) r(x_0) \, dx_0$$

To show equation (i), note that we have a continuous-state process; we proceed as follows. By definition (for $k \geq 1$) $P_k(t + \Delta t, x_0 + \Delta t) \Delta x_0$ is the probability that, at time $t + \Delta t$, there are k in system and the customer being served has already received $x_0 + \Delta t$ seconds of service. The only two ways to enter this state are:

(i) there were k in system at time t, the customer in service had received x_0 seconds of service at time t, and there were no arrivals and no departures in $(t, t + \Delta t)$; or

(ii) there were $k - 1$ in system at time t, the customer in service had received x_0 seconds of service at time t, and there was one arrival in $(t, t + \Delta t)$ but no departures in $(t, t + \Delta t)$.

Thus

$$P_k(t + \Delta t, x_0 + \Delta t) \Delta x_0$$

$$= [1 - \lambda \Delta t + o(\Delta t)][1 - r(x_0) \Delta t + o(\Delta t)] P_k(t, x_0) \Delta x_0$$

$$+ [\lambda \Delta t + o(\Delta t)][1 - r(x_0) \Delta t + o(\Delta t)] P_{k-1}(t, x_0) \Delta x_0$$

Expanding $P_k(t + \Delta t, x_0 + \Delta t)$ as

$$P_k(t + \Delta t, x_0 + \Delta t) = P_k(t, x_0 + \Delta t) + \frac{\partial P_k(t, x_0 + \Delta t)}{\partial t} \Delta t + o(\Delta t)$$

we have the equation

$$P_k(t, x_0 + \Delta t) - P_k(t, x_0) + \frac{\partial P_k(t, x_0 + \Delta t)}{\partial t} \Delta t$$

$$= -[\lambda + r(x_0)] P_k(t, x_0) \Delta t + \lambda P_{k-1}(t, x_0) \Delta t + o(\Delta t)$$

Dividing by Δt and letting $\Delta t \to 0$ gives

$$\frac{\partial P_k(t, x_0)}{\partial x_0} + \frac{\partial P_k(t, x_0)}{\partial t} = -[\lambda + r(x_0)] P_k(t, x_0) + \lambda P_{k-1}(t, x_0)$$

Letting $t \to \infty$ and using the equilibrium result $\lim_{t \to \infty} \partial P_k(t, x_0)/\partial t = 0$ we finally obtain

$$\frac{\partial p_k(x_0)}{\partial x_0} = -[\lambda + r(x_0)] p_k(x_0) + \lambda p_{k-1}(x_0) \qquad (k \geq 1)$$

which is equation (i).

We obtain equations (ii) and (iii) as follows. Consider the state in which there are $k \geq 1$ in the system at time $t + \Delta t$ and the customer in service has received Δt seconds of service. For $k > 1$, the only way to enter this state is to have $k + 1$ at time t, no arrivals in $(t, t + \Delta t)$, and one departure in $(t, t + \Delta t)$. For $k = 1$ there is the added possibility of having an empty system at time t, but one arrival in $(t, t + \Delta t)$. For $k > 1$, we have [conditioning on $X_0(t) = x_0$]

$$P_k(t + \Delta t, 0) \Delta t = \int_0^\infty [1 - \lambda \Delta t + o(\Delta t)][r(x_0) \Delta t + o(\Delta t)]P_{k+1}(t, x_0) \, dx_0$$

$$= \int_0^\infty r(x_0)P_{k+1}(t, x_0) \, dx_0 \, \Delta t + o(\Delta t)$$

Dividing by Δt and letting $\Delta t \to 0$ gives

$$P_k(t, 0) = \int_0^\infty P_{k+1}(t, x_0)r(x_0) \, dx_0$$

Letting $t \to \infty$, we finally have

$$p_k(0) = \int_0^\infty p_{k+1}(x_0)r(x_0) \, dx_0 \qquad (k > 1)$$

For $k = 1$, a similar argument (we may now have an arrival to an empty system) gives

$$p_1(0) = \int_0^\infty p_1(x_0)r(x_0) \, dx_0 + \lambda p_0$$

(c) For $k \geq 1$, we clearly have $p_k = \int_0^\infty p_k(x_0) \, dx_0$. Thus the normalization $\sum_{k=0}^\infty p_k = 1$ becomes

$$p_0 + \sum_{k=1}^\infty \int_0^\infty p_k(x_0) \, dx_0 = 1$$

(d) Multiply each equation in (i) of part (b) by z^k and sum for $k \geq 1$. Then

$$\sum_{k=1}^\infty \frac{\partial p_k(x_0)}{\partial x_0} z^k = -[\lambda + r(x_0)] \sum_{k=1}^\infty p_k(x_0)z^k + \lambda \sum_{k=1}^\infty p_{k-1}(x_0)z^k$$

Recalling that $p_0(x_0) = 0$ and $R(z, x_0) = \sum_{k=1}^\infty p_k(x_0)z^k$ we find

$$\frac{\partial R(z, x_0)}{\partial x_0} = [\lambda z - \lambda - r(x_0)]R(z, x_0)$$

Next, using (ii) and (iii) of part (b),

$$R(z,0) = \sum_{k=1}^{\infty} p_k(0) z^k$$

$$= \lambda p_0 z + \sum_{k=1}^{\infty} \int_0^{\infty} p_{k+1}(x_0) r(x_0) \, dx_0 \, z^k$$

$$= \lambda p_0 z + \int_0^{\infty} \left[\sum_{k=1}^{\infty} p_{k+1}(x_0) z^k \right] r(x_0) \, dx_0$$

Using the definition of $R(z, x_0)$ we finally have

$$R(z,0) = \lambda p_0 z + \int_0^{\infty} \frac{1}{z} [R(z, x_0) - p_1(x_0) z] r(x_0) \, dx_0$$

$$= \lambda p_0 z + \frac{1}{z} \int_0^{\infty} R(z, x_0) r(x_0) \, dx_0 - \lambda p_0$$

or

$$zR(z,0) = \lambda p_0 z(z - 1) + \int_0^{\infty} R(z, x_0) r(x_0) \, dx_0$$

(e) From part (d) we have

$$\frac{\dfrac{\partial R(z, y)}{\partial y}}{R(z, y)} = \lambda z - \lambda - r(y)$$

Integrate from 0 to x_0 to obtain

$$\log_e[R(z, x_0)] - \log_e[R(z, 0)] = [\lambda z - \lambda] x_0 - \int_0^{x_0} r(y) \, dy$$

or

$$R(z, x_0) = R(z, 0) e^{\lambda(z-1)x_0 - \int_0^{x_0} r(y) \, dy}$$

as desired. We next note that

$$e^{- \int_0^{x_0} r(y) \, dy} = e^{- \int_0^{x_0} \frac{b(y)}{1 - B(y)} \, dy} = e^{\log_e[1 - B(x_0)]} = 1 - B(x_0)$$

So we can rewrite our previous result as

$$R(z, x_0) = R(z, 0) e^{\lambda(z-1)x_0} [1 - B(x_0)]$$

Using this expression in the equation for $zR(z, 0)$ obtained in part (d) yields

$$zR(z, 0) = \lambda p_0 z(z - 1) + \int_0^\infty R(z, 0)e^{\lambda(z-1)x_0}[1 - B(x_0)]r(x_0)\,dx_0$$

or

$$zR(z, 0) = \lambda p_0 z(z - 1) + R(z, 0)\int_0^\infty e^{\lambda(z-1)x_0}b(x_0)\,dx_0$$

$$= \lambda p_0 z(z - 1) + R(z, 0)B^*(\lambda - \lambda z)$$

Thus

$$R(z, 0) = \frac{\lambda z(z - 1)p_0}{z - B^*(\lambda - \lambda z)}$$

(f) We have

$$R(z) \overset{\Delta}{=} \int_0^\infty R(z, x_0)\,dx_0$$

$$= \int_0^\infty R(z, 0)e^{\lambda(z-1)x_0}[1 - B(x_0)]\,dx_0$$

$$= R(z, 0)\int_0^\infty e^{-\lambda(1-z)x_0}[1 - B(x_0)]\,dx_0$$

Integrating gives

$$R(z) = R(z, 0)\frac{1 - B^*(\lambda - \lambda z)}{\lambda(1 - z)}$$

(g) From the normalization equation $p_0 + \int_0^\infty \sum_{k=1}^\infty p_k(x_0)\,dx_0 = 1$ and the definition of $R(1, x_0)$, we have $p_0 + \int_0^\infty R(1, x_0)\,dx_0 = 1$. Thus

$$p_0 + R(1) = 1$$

From (f) and (e) we also have

$$R(z) = R(z, 0)\frac{1 - B^*(\lambda - \lambda z)}{\lambda(1 - z)}$$

$$= \frac{\lambda z(z - 1)p_0}{z - B^*(\lambda - \lambda z)} \cdot \frac{1 - B^*(\lambda - \lambda z)}{\lambda(1 - z)}$$

or

$$\frac{R(z)}{z} = p_0\frac{1 - B^*(\lambda - \lambda z)}{B^*(\lambda - \lambda z) - z}$$

Letting $z \to 1$ (and using L'Hospital's rule) we find

$$R(1) = p_0 \lim_{z \to 1} \frac{1 - B^*(\lambda - \lambda z)}{B^*(\lambda - \lambda z) - z}$$

$$= p_0 \left. \frac{-B^{*(1)}(\lambda - \lambda z)(-\lambda)}{B^{*(1)}(\lambda - \lambda z)(-\lambda) - 1} \right|_{z=1}$$

$$= p_0 \frac{-\lambda \bar{x}}{\lambda \bar{x} - 1} = p_0 \frac{\rho}{1 - \rho}$$

But then $1 = p_0 + R(1) = p_0 + p_0\rho/(1 - \rho)$, and so

$$p_0 = 1 - \rho$$

(h) We have

$$Q(z) = p_0 + R(z)$$

$$= p_0 + \frac{p_0 z[1 - B^*(\lambda - \lambda z)]}{B^*(\lambda - \lambda z) - z}$$

$$= p_0 \frac{B^*(\lambda - \lambda z)(1 - z)}{B^*(\lambda - \lambda z) - z}$$

So

$$Q(z) = B^*(\lambda - \lambda z)\frac{(1 - \rho)(1 - z)}{B^*(\lambda - \lambda z) - z}$$

which is Eq. (1.104).   □

PROBLEM 5.8

Consider the M/G/∞ queue in which each customer always finds a free server; thus $s(y) = b(y)$ and $T = \bar{x}$. Let $P_k(t) = P[N(t) = k]$ and assume $P_0(0) = 1$.

(a) Show that

$$P_k(t) = \sum_{n=k}^{\infty} e^{-\lambda t}\frac{(\lambda t)^n}{n!}\binom{n}{k}\left[\frac{1}{t}\int_0^t [1 - B(x)]\,dx\right]^k\left[\frac{1}{t}\int_0^t B(x)\,dx\right]^{n-k}$$

[HINT: $(1/t)\int_0^t B(x)\,dx$ is the probability that a customer's service terminates by time t, given that his arrival time was uniformly distributed over the interval $(0, t)$.]

(b) Show that $p_k \triangleq \lim P_k(t)$ as $t \to \infty$ is

$$p_k = \frac{(\lambda \bar{x})^k}{k!} e^{-\lambda \bar{x}}$$

regardless of the form of $B(x)$!

SOLUTION

(a) We determine the distribution of $N(t)$, the number of customers in system at time t, by conditioning on the number of arrivals in $(0, t)$. Thus

$$P_k(t) = P[N(t) = k]$$

$$= \sum_{n=0}^{\infty} P[N(t) = k \mid n \text{ arrivals in } (0, t)] P[n \text{ arrivals in } (0, t)]$$

$$= \sum_{n=0}^{\infty} P[N(t) = k \mid n \text{ arrivals in } (0, t)] \frac{e^{-\lambda t}(\lambda t)^n}{n!}$$

The probability that a customer who arrived at time x $(0 < x < t)$ is still in the system at time t is simply $1 - B(t - x)$. Since arrivals form a Poisson process, the joint distribution of the arrival times given that n arrivals occurred in $(0, t)$ is the same as the distribution of n points uniformly distributed over $(0, t)$. Thus for any number n of arrivals in $(0, t)$, the probability of any customer still being present at time t is

$$\int_0^t [1 - B(t - x)] \frac{dx}{t} = \frac{1}{t} \int_0^t [1 - B(x)] dx$$

Hence the probability that a customer who arrived in $(0, t)$ will have completed service by time t is

$$1 - \frac{1}{t} \int_0^t [1 - B(x)] dx = \frac{1}{t} \int_0^t B(x) dx$$

Thus

$$P[N(t) = k \mid n \text{ arrivals in } (0, t)]$$

$$= \binom{n}{k} \left[\frac{1}{t} \int_0^t [1 - B(x)] dx \right]^k \left[\frac{1}{t} \int_0^t B(x) dx \right]^{n-k}$$

for $k \le n$ (and $= 0$ for $k > n$). So

$$P_k(t) = \sum_{n=k}^{\infty} \binom{n}{k} \left[\frac{1}{t} \int_0^t [1 - B(x)] dx \right]^k \left[\frac{1}{t} \int_0^t B(x) dx \right]^{n-k} \frac{e^{-\lambda t}(\lambda t)^n}{n!}$$

which is the desired result.

(b) We first simplify the expression obtained in part (a), and then find the limit as $t \to \infty$.

$$P[N(t) = k] = \sum_{n=k}^{\infty} e^{-\lambda t} \frac{(\lambda t)^n}{n!} \binom{n}{k} \left[\frac{1}{t} \int_0^t [1 - B(x)] \, dx \right]^k \left[\frac{1}{t} \int_0^t B(x) \, dx \right]^{n-k}$$

$$= e^{-\lambda t} \sum_{n=k}^{\infty} \frac{\left[\lambda \int_0^t [1 - B(x)] \, dx \right]^k}{k!} \cdot \frac{\left[\lambda \int_0^t B(x) \, dx \right]^{n-k}}{(n-k)!}$$

$$= e^{-\lambda t} \frac{\left[\lambda \int_0^t [1 - B(x)] \, dx \right]^k}{k!} \sum_{n=0}^{\infty} \frac{\left[\lambda \int_0^t B(x) \, dx \right]^n}{n!}$$

$$P[N(t) = k] = e^{-\lambda t} \frac{\left[\lambda \int_0^t [1 - B(x)] \, dx \right]^k}{k!} e^{\lambda \int_0^t B(x) \, dx}$$

$$= e^{-\lambda \int_0^t [1 - B(x)] \, dx} \frac{\left[\lambda \int_0^t [1 - B(x)] \, dx \right]^k}{k!}$$

Thus, *for every* t, $N(t)$ is Poisson with parameter $\lambda \int_0^t [1 - B(x)] \, dx$. Letting $t \to \infty$ and noting that

$$\lim_{t \to \infty} \int_0^t [1 - B(x)] \, dx = \int_0^\infty [1 - B(x)] \, dx = \bar{x}$$

we see immediately that

$$p_k \overset{\Delta}{=} \lim_{t \to \infty} P_k(t) = e^{-\lambda \bar{x}} \frac{(\lambda \bar{x})^k}{k!}$$

Thus as $t \to \infty$, the limiting distribution of number in system is Poisson with parameter $\lambda \bar{x}$, which is independent (except for the mean) of $B(x)$.   □

PROBLEM 5.9

Consider M/E$_2$/1.

(a) Find the polynomial for $G^*(s)$.
(b) Solve for $S(y) = P[\text{time in system} \le y]$.

SOLUTION

(a) For the M/E$_2$/1 system, the Laplace transform of the service time density is

$$B^*(s) = \left(\frac{2\mu}{s + 2\mu} \right)^2$$

Thus Eq. (1.111) gives

$$G^*(s) = \left[\frac{2\mu}{s + \lambda - \lambda G^*(s) + 2\mu} \right]^2$$

Expanding, we get

$$\lambda^2[G^*(s)]^3 - 2\lambda(s + \lambda + 2\mu)[G^*(s)]^2 + (s + \lambda + 2\mu)^2 G^*(s) - 4\mu^2 = 0$$

(b) Equation (1.106) gives

$$S^*(s) = B^*(s) \frac{s(1 - \rho)}{s - \lambda + \lambda B^*(s)}$$

Thus

$$S^*(s) = \left(\frac{2\mu}{s + 2\mu} \right)^2 \frac{s(1 - \rho)}{s - \lambda + \lambda \left(\dfrac{2\mu}{s + 2\mu} \right)^2}$$

$$= \frac{4\mu^2(1 - \rho)}{s^2 + (4\mu - \lambda)s + 4\mu(\mu - \lambda)}$$

The denominator $s^2 + (4\mu - \lambda)s + 4\mu(\mu - \lambda)$ has roots s_1, s_2 (where $\rho = \lambda/\mu$):

$$s_1 = \frac{-\mu(4 - \rho) + \mu\sqrt{\rho^2 + 8\rho}}{2}$$

$$s_2 = \frac{-\mu(4 - \rho) - \mu\sqrt{\rho^2 + 8\rho}}{2}$$

We note that, for $\rho < 1$, we have $16\rho < 16$ and thus $(4 - \rho)^2 > \rho^2 + 8\rho$. Hence $s_2 < s_1 < 0$ for $0 < \rho < 1$. Factoring,

$$S^*(s) = \frac{4\mu^2(1 - \rho)}{(s - s_1)(s - s_2)}$$

$$= \frac{4\mu^2(1 - \rho)}{\mu\sqrt{\rho^2 + 8\rho}} \left(\frac{1}{s - s_1} - \frac{1}{s - s_2} \right)$$

Invert to find the pdf $s(y)$ as

$$s(y) = \frac{4\mu(1 - \rho)}{\sqrt{\rho^2 + 8\rho}} (e^{s_1 y} - e^{s_2 y})$$

Thus the PDF $S(y)$ is

$$S(y) = \frac{4\mu(1 - \rho)}{\sqrt{\rho^2 + 8\rho}} \left[\frac{1}{s_1} (e^{s_1 y} - 1) - \frac{1}{s_2} (e^{s_2 y} - 1) \right] \qquad \square$$

PROBLEM 5.10

Consider an M/D/1 system for which $\bar{x} = 2$ sec.

(a) Show that the residual service time pdf $\hat{b}(x)$ is a rectangular distribution.
(b) For $\rho = 0.25$, show that the result of Eq. (1.108) with four terms may be used as a good approximation to the distribution of queueing time.

SOLUTION

(a) The service time distribution is given by

$$B(x) = \begin{cases} 0 & x < 2 \\ 1 & x \geq 2 \end{cases}$$

The residual service time pdf is

$$\hat{b}(x) = \frac{1 - B(x)}{\bar{x}} = \begin{cases} \frac{1}{2} & x < 2 \\ 0 & x \geq 2 \end{cases}$$

Thus $\hat{b}(x)$ is rectangular.

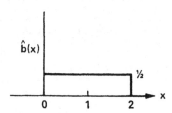

(b) The first four terms of the series in Eq. (1.108) give

$$w(y) \cong w_{\text{approx}}(y) \triangleq (1 - \rho)\left[u_0(y) + \rho\hat{b}(y) + \rho^2\hat{b}_{(2)}(y) + \rho^3\hat{b}_{(3)}(y)\right]$$

As

$$\hat{b}(y) = \begin{cases} \frac{1}{2} & y < 2 \\ 0 & y \geq 2 \end{cases}$$

we see that

$$\hat{b}_{(2)}(y) = \begin{cases} \dfrac{y}{4} & 0 \leq y \leq 2 \\[2mm] 1 - \dfrac{y}{4} & 2 \leq y \leq 4 \end{cases}$$

and

$$
\hat{b}_{(3)}(y) = \begin{cases} \dfrac{y^2}{16} & 0 \le y \le 2 \\[2ex] \dfrac{-y^2}{8} + \dfrac{3}{4}y - \dfrac{3}{4} & 2 \le y \le 4 \\[2ex] \dfrac{y^2}{16} - \dfrac{3}{4}y + \dfrac{9}{4} & 4 \le y \le 6 \end{cases}
$$

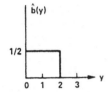
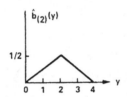
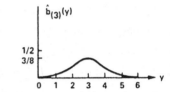

We compare $w(y)$ and $w_{\text{approx}}(y)$ in three different ways.

First, the area A under the curve $w(y)$ minus the area A_{approx} under the curve $w_{\text{approx}}(y)$ is

$$
A - A_{\text{approx}} = (1 - \rho) \sum_{k=4}^{\infty} \rho^k \int_0^{\infty} \hat{b}_{(k)}(y)\, dy = (1 - \rho) \sum_{k=4}^{\infty} \rho^k
$$

$$
= (1 - \rho)\rho^4 \left(\frac{1}{1 - \rho} \right) = \rho^4
$$

As $\rho = \frac{1}{4}$,

$$
A - A_{\text{approx}} = \tfrac{1}{256}
$$

Thus, in terms of area, we have a "good" approximation.

Second, we note that $w_{\text{approx}}(y) = 0$ for $y \ge 6$. Thus the tail of the density $w(y)$ is *not* approximated very well.

Third, we compare the mean wait W with an approximation W_{approx} calculated from $w_{\text{approx}}(y)$. [Note that $w_{\text{approx}}(y)$ is not a pdf.]

$$
W = \int_0^{\infty} y w(y)\, dy = (1 - \rho) \sum_{k=1}^{\infty} \rho^k \int_0^{\infty} y \hat{b}_{(k)}(y)\, dy
$$

We now observe that $\int_0^{\infty} y \hat{b}_{(k)}(y)\, dy$ has value k, since it represents the mean of a sum of k random variables each having mean 1. Thus

$$W = (1 - \rho) \sum_{k=1}^{\infty} k\rho^k = (1 - \rho)\rho \frac{\partial}{\partial \rho} \left(\frac{1}{1 - \rho} \right)$$

or

$$W = \frac{\rho}{1 - \rho}$$

For $\rho = \frac{1}{4}$,

$$W = \frac{1}{3}$$

Now

$$W_{approx} = \int_0^{\infty} y w_{approx}(y) \, dy = (1 - \rho) \sum_{k=1}^{3} \rho^k \int_0^{\infty} y \hat{b}_{(k)}(y) \, dy$$

$$= (1 - \rho) \sum_{k=1}^{3} k\rho^k = (1 - \rho)(\rho + 2\rho^2 + 3\rho^3)$$

For $\rho = \frac{1}{4}$,

$$W_{approx} = \frac{3}{4} \left(\frac{1}{4} + \frac{1}{8} + \frac{3}{64} \right) = \frac{3}{4} \cdot \frac{27}{64} = \frac{81}{256}$$

or

$$W_{approx} = 0.31640625$$

Thus

$$\frac{W - W_{approx}}{W} = \frac{\frac{1}{3} - \left(\frac{3}{4} \right)^4}{\frac{1}{3}} \cong 0.0508$$

and so W_{approx} is within 5% of the mean W.   □

PROBLEM 5.11

Consider an M/G/1 queue in which bulk arrivals occur at rate λ and with a probability g_r that r customers arrive together at an arrival instant.

(a) Show that the z-transform of the number of customers arriving in an interval of length t is $e^{-\lambda t[1 - G(z)]}$, where $G(z) = \sum g_r z^r$.
(b) Show that the z-transform of the random variables $\tilde{v}$, the number of arrivals during the service of a customer, is $B^*(\lambda - \lambda G(z))$.

SOLUTION

(a) Let the random variable $N = N(t)$ represent the number of bulks that arrive in an interval of length t, and let $N_1(z)$ be its z-transform. The total number of customers to arrive is $Y = X_1 + \cdots + X_N$, where X_i is the size of the ith bulk. To find the z-transform $Y(z)$ for number of customers, we proceed by conditioning on N as follows:

$$Y(z \mid N = n) \overset{\Delta}{=} E[z^Y \mid N = n] = E[z^{X_1 + \cdots + X_n}]$$
$$= E[z^{X_1}] \cdots E[z^{X_n}] = [G(z)]^n$$

Unconditioning gives

$$Y(z) = \sum_{n=0}^{\infty} Y(z \mid N = n) P[N = n]$$

$$= \sum_{n=0}^{\infty} [G(z)]^n P[N = n] = N_1(G(z))$$

But $N_1(z) = e^{-\lambda t(1-z)}$ since $N = N(t)$ is Poisson with parameter λt. Thus we have

$$Y(z) = N_1(G(z)) = e^{-\lambda t[1 - G(z)]}$$

(b) We have

$$V(z) = E[z^{\tilde{v}}] = \sum_{k=0}^{\infty} P[\tilde{v} = k] z^k$$

$$= \sum_{k=0}^{\infty} \int_0^{\infty} P[\tilde{v} = k \mid \tilde{x} = x] b(x) \, dx \, z^k$$

$$= \int_0^{\infty} \sum_{k=0}^{\infty} P[\tilde{v} = k \mid \tilde{x} = x] z^k b(x) \, dx$$

By part (a) above, $\sum_{k=0}^{\infty} P[\tilde{v} = k \mid \tilde{x} = x] z^k = e^{-\lambda x[1 - G(z)]}$. (It is the z-transform of the number of customers to arrive during an interval of length x.) Thus

$$V(z) = \int_0^{\infty} e^{-\lambda x[1 - G(z)]} b(x) \, dx$$

or

$$V(z) = B^*(\lambda - \lambda G(z))$$

In the case of single arrivals (i.e., no bulks), we have that $g_1 = 1$ and $g_r = 0$ for $r \neq 1$. Then $G(z) = z$, and so

$$V(z) = B^*(\lambda - \lambda z)$$

As an additional insight, we note that this equation relates the z-transform of the number of arrivals from a Poisson process at rate λ to the Laplace transform of the length of the time interval over which those arrivals occur; thus such a relationship applies to any interval during which Poisson arrivals are counted. □

PROBLEM 5.12

Consider the M/G/1 bulk arrival system in the previous problem. Using the method of imbedded Markov chains:

(a) Find the expected queue size at departure instants. [HINT: show that $\bar{v} = \rho = \lambda \bar{x} \bar{g}$ and

$$\overline{v^2} - \bar{v} = \left. \frac{d^2 V(z)}{dz^2} \right|_{z=1} = \rho^2(1 + C_b^2) + \rho\bar{g}(1 + C_g^2) - \rho$$

where C_g is the coefficient of variation of the bulk group size and $\bar{g}$ is the mean group size.]

(b) Show that the generating function for queue size at departure instants is

$$Q(z) = \frac{(1 - \rho)[1 - G(z)]B^*(\lambda - \lambda G(z))}{\bar{g}[B^*(\lambda - \lambda G(z)) - z]}$$

(c) Using the same method (imbedded Markov chains) find the expected number of groups in the queue averaged over departure times. [HINTS: Show that $D(z) = \beta^*(\lambda - \lambda z)$, where $D(z)$ is the generating function for the number of *groups* arriving during the service time for an entire *group* and where $\beta^*(s)$ is the Laplace transform of the service time density for an entire group. Also note that $\beta^*(s) = G(B^*(s))$, which allows us to show that $\overline{\tau^2} = (\bar{x})^2(\overline{g^2} - \bar{g}) + \overline{x^2}\bar{g}$, where $\overline{\tau^2}$ is the second moment of the *group* service time.]

(d) Using Little's result, find W_g, the expected wait on queue for a group (measured from the arrival time of the group until the start of service of the *first* member of the group) and show that

$$W_g = \frac{\rho \bar{x} \bar{g}}{2(1 - \rho)} \left[1 + \frac{C_b^2}{\bar{g}} + C_g^2\right]$$

(e) If the customers within a group arriving together are served in random order, show that the ratio of the mean waiting time for a single customer to the average service time for a single customer is $W_g/\bar{x}$ from (d) increased by $\frac{1}{2}\bar{g}(1 + C_g^2) - \frac{1}{2}$.

SOLUTION

(a) We begin by finding the first two moments of the random variable $\tilde{v}$ (which has the same distribution as v_n) as suggested in the hint. By Problem 5.11(b), $V(z) = B^*(\lambda - \lambda G(z))$. Thus

$$V^{(1)}(z) = B^{*(1)}(\lambda - \lambda G(z))[-\lambda G^{(1)}(z)]$$

and

$$V^{(2)}(z) = B^{*(2)}(\lambda - \lambda G(z))[\lambda G^{(1)}(z)]^2 + B^{*(1)}(\lambda - \lambda G(z))[-\lambda G^{(2)}(z)]$$

Hence

$$\bar{v} = V^{(1)}(1) = B^{*(1)}(0)[-\lambda G^{(1)}(1)] = \lambda \bar{g} \bar{x}$$

Since the average arrival rate is $\lambda \bar{g}$, we have

$$\bar{v} = \lambda \bar{g} \bar{x} = \rho$$

Also

$$\overline{v^2} - \bar{v} = V^{(2)}(1) = B^{*(2)}(0)[\lambda G^{(1)}(1)]^2 + B^{*(1)}(0)[-\lambda G^{(2)}(1)]$$

$$= \overline{x^2}(\lambda \bar{g})^2 + \lambda \bar{x}(\overline{g^2} - \bar{g})$$

$$= (\lambda \bar{g} \bar{x})^2 \frac{\overline{x^2}}{(\bar{x})^2} + \lambda \bar{x} \bar{g} \left[\frac{\overline{g^2}}{\bar{g}} - 1 \right]$$

$$= \rho^2(1 + C_b^2) + \rho \bar{g}(1 + C_g^2) - \rho$$

Thus

$$\overline{v^2} = \rho^2(1 + C_b^2) + \rho \bar{g}(1 + C_g^2)$$

We next derive an equation for q_{n+1}, the number left behind by a departure. We first note that if $q_n > 0$, then clearly $q_{n+1} = q_n - 1 + v_{n+1}$. From the previous problem, we know that $\tilde{v}$ has the z-transform $V(z) = B^*(\lambda - \lambda G(z))$. However, if $q_n = 0$ the next customer to arrive to this empty system will be in a bulk whose size is a random variable, say, $\tilde{g}$; thus $q_{n+1} = \tilde{g} - 1 + v_{n+1}$. Hence

$$q_{n+1} = \begin{cases} v_{n+1} - 1 + q_n & q_n > 0 \\ v_{n+1} - 1 + \tilde{g} & q_n = 0 \end{cases}$$

where $\tilde{g}$ has z-transform $G(z)$. Define a new random variable

$$\Delta_{\tilde{g},k} \overset{\Delta}{=} \begin{cases} k & k > 0 \\ \tilde{g} & k = 0 \end{cases}$$

Thus we have $q_{n+1} = v_{n+1} - 1 + \Delta_{\tilde{g}.q_n}$. Take expectations and let $n \to \infty$.
Then $\bar{q} = \bar{v} - 1 + E[\Delta_{\tilde{g}.\tilde{q}}]$. But

$$E[\Delta_{\tilde{g}.\tilde{q}}] = E[\tilde{g}]P[\tilde{q} = 0] + \sum_{k=1}^{\infty} kP[\tilde{q} = k]$$

and so

$$E[\Delta_{\tilde{g}.\tilde{q}}] = \bar{g}d_0 + \bar{q}$$

Therefore

$$\bar{q} = \bar{v} - 1 + \bar{g}d_0 + \bar{q}$$

or

$$d_0 = \frac{1 - \bar{v}}{\bar{g}} = \frac{1 - \rho}{\bar{g}}$$

To find $\bar{q}$, we must square our equation and try again.

$$q_{n+1}^2 = (v_{n+1} - 1)^2 + 2(v_{n+1} - 1)\Delta_{\tilde{g}.q_n} + \Delta_{\tilde{g}.q_n}^2$$

Taking expectations, using independence, and letting $n \to \infty$, we have

$$\overline{q^2} = \overline{v^2} - 2\bar{v} + 1 + 2(\bar{v} - 1)E[\Delta_{\tilde{g}.\tilde{q}}] + E[\Delta_{\tilde{g}.\tilde{q}}^2]$$

But

$$E[\Delta_{\tilde{g}.\tilde{q}}^2] = E[\tilde{g}^2]P[\tilde{q} = 0] + \sum_{k=1}^{\infty} k^2 P[\tilde{q} = k]$$

$$= \overline{g^2}d_0 + \overline{q^2}$$

So

$$\overline{q^2} = \overline{v^2} - 2\bar{v} + 1 + 2(\bar{v} - 1)(\bar{g}d_0 + \bar{q}) + \overline{g^2}d_0 + \overline{q^2}$$

Substituting for $\overline{v^2}$, $\bar{v}$, and d_0 we have

$$0 = \rho^2(1 + C_b^2) + \rho\bar{g}(1 + C_g^2) - 2\rho + 1 + 2(\rho - 1)(1 - \rho + \bar{q}) + \frac{\overline{g^2}}{\bar{g}}(1 - \rho)$$

or

$$2(1 - \rho)\bar{q} = \rho^2(1 + C_b^2) + \rho\bar{g}(1 + C_g^2) + 2(1 - \rho) - 1 - 2(1 - \rho)^2$$
$$+ \bar{g}(1 + C_g^2)(1 - \rho)$$

and so

$$2(1 - \rho)\bar{q} = \rho^2(1 + C_b^2) + \bar{g}(1 + C_g^2) + 2(1 - \rho)\rho - 1$$

Thus

$$\bar{q} = \rho + \frac{\rho^2(1 + C_b^2)}{2(1 - \rho)} + \frac{\bar{g}(1 + C_g^2) - 1}{2(1 - \rho)}$$

(b) We have

$$Q_{n+1}(z) = E\left[z^{q_{n+1}}\right] = E\left[z^{v_{n+1}-1+\Delta_{\tilde{g},q_n}}\right]$$

$$= E\left[z^{v_{n+1}-1}\right]E\left[z^{\Delta_{\tilde{g},q_n}}\right]$$

by independence of v_{n+1}, $\tilde{g}$, q_n. Let $n \to \infty$ and obtain

$$Q(z) = \frac{V(z)}{z}E\left[z^{\Delta_{\tilde{g},\tilde{q}}}\right]$$

But

$$E\left[z^{\Delta_{\tilde{g},\tilde{q}}}\right] = E[z^{\tilde{g}}]P[\tilde{q} = 0] + \sum_{k=1}^{\infty} z^k P[\tilde{q} = k]$$

$$= G(z)d_0 + [Q(z) - d_0]$$

Thus

$$Q(z) = \frac{V(z)}{z}[Q(z) + [G(z) - 1]d_0]$$

$$Q(z) = V(z)\frac{d_0[1 - G(z)]}{V(z) - z}$$

But $V(z) = B^*(\lambda - \lambda G(z))$ and $d_0 = (1 - \rho)/\bar{g}$. Thus

$$Q(z) = B^*(\lambda - \lambda G(z))\frac{(1 - \rho)[1 - G(z)]}{\bar{g}[B^*(\lambda - \lambda G(z)) - z]}$$

(c) Since we seek the expected number of groups in the queue averaged over departure instants, we study an M/G/1 queue where the nth group is considered to be the nth "customer" whose service time corresponds to that for the entire nth group. Recognizing this, we may apply the M/G/1 results directly; first, however, the arrival and service processes must be determined. From Problem 5.11, we have $D(z) = \beta^*(\lambda - \lambda z)$. Similar to the development in part (a) of the previous problem, we find that $\beta^*(s) = G(B^*(s))$. (The nth group's service time is composed of a sum of a random number—the bulk size—of individual service times.) So by Eq. (1.102) and Eq. (1.36) we have:

$$\overline{D}_q \stackrel{\Delta}{=} \text{expected number of groups in queue}$$

satisfies

$$\overline{D}_q = \frac{\lambda^2\overline{\tau^2}}{2(1 - \lambda\overline{\tau})}$$

where $\overline{\tau}$ and $\overline{\tau^2}$ are the first two moments of the group service time. Note that

$$\overline{\tau} = -\beta^{*(1)}(0) = \bar{g}\,\bar{x}$$

(the perfectly reasonable result that the mean group service time is just $\bar{x}$ times the mean bulk size) and

$$\overline{\tau^2} = \beta^{*(2)}(0) = \overline{g}\,\overline{x^2} + (\overline{g^2} - \overline{g})(\overline{x})^2$$

We may use these results and the fact that $\rho = \lambda\bar{\tau}$ to obtain

$$\overline{D}_q = \frac{\lambda^2[\overline{g}\,\overline{x^2} + (\overline{g^2} - \overline{g})(\overline{x})^2]}{2(1 - \rho)}$$

$$\overline{D}_q = \frac{(\lambda\,\overline{g}\,\overline{x})^2}{2(1 - \rho)}\left[\frac{\overline{g^2}}{\overline{g}^2} + \left(\frac{\overline{x^2} - (\overline{x})^2}{\overline{g}(\overline{x})^2}\right)\right]$$

$$\overline{D}_q = \frac{\rho^2}{2(1 - \rho)}\left[1 + C_g^2 + \frac{C_b^2}{\overline{g}}\right]$$

(d) By Little's result (applied on the "group" queue)

$$\overline{D}_q = \lambda W_g$$

Hence

$$W_g = \frac{\rho\,\overline{g}\,\overline{x}}{2(1 - \rho)}\left[1 + C_g^2 + \frac{C_b^2}{\overline{g}}\right]$$

(e) The waiting time for a customer consists of the waiting time until someone in his group gets served [expected value W_g was found in part (d)] plus the waiting time due to service of members within his group (call the expected value of this waiting time W_s). Pick a random customer. Given that this customer came from a bulk of size r, his expected wait due to people in this bulk is

$$\frac{1}{r}[0 + \bar{x} + 2\bar{x} + \cdots + (r - 1)\bar{x}] = \frac{1}{r}\cdot\frac{r(r - 1)}{2}\bar{x} = \frac{r - 1}{2}\bar{x}$$

We now must find the probability, say, $\hat{p}_r$, that this random customer came from a bulk of size r. To this end, consider an arbitrarily long time interval u. Bulks of size r arrive at a rate λg_r, each bringing r customers. Thus an average of $\lambda g_r r u$ customers from bulks of size r will arrive during u. Similarly, the total number of customers to arrive during u will be, on the average, simply $\sum_{k=1}^{\infty} \lambda g_k k u$. Thus the probability $\hat{p}_r$ that a random customer will be in a bulk of size r is approximated by

$$\hat{p}_r \cong \frac{\lambda g_r r u}{\sum_{k=1}^{\infty} \lambda g_k k u}$$

Letting $u \to \infty$ we obtain

$$\hat{p}_r = P[\text{random customer was in bulk of size } r]$$

$$= \frac{r g_r}{\sum_{k=1}^{\infty} k g_k} = \frac{r g_r}{\overline{g}}$$

(Note: This type of argument extends to the continuous variable case.) So

$$W_s = \sum_{r=1}^{\infty} (W_s \mid \text{customer from bulk of size } r)\hat{p}_r$$

$$= \sum_{r=1}^{\infty} \left(\frac{r-1}{2} \bar{x} \right) \frac{r g_r}{\bar{g}}$$

$$W_s = \frac{\bar{x}}{2\bar{g}} \left[\sum_{r=1}^{\infty} (r^2 - r)g_r \right] = \frac{\bar{x}}{2\bar{g}} \left[\overline{g^2} - \bar{g} \right]$$

and thus the normalized increase is

$$\frac{W_s}{\bar{x}} = \frac{1}{2} \bar{g} \left[\frac{\overline{g^2}}{(\bar{g})^2} \right] - \frac{1}{2} = \frac{1}{2}\bar{g}(1 + C_g^2) - \frac{1}{2}$$

[Note: A second method of proof for part (e) (using Little's result) is as follows: We first appeal to a known result to obtain the distribution of number in system for this bulk queue over all time. To this end define

$$P_k(t) = P[k \text{ in system at time } t]$$

and let

$$p_k = \lim_{t \to \infty} P_k(t) \qquad \text{for } k = 0, 1, 2, \ldots$$

Define the z-transform $P(z) \stackrel{\Delta}{=} \sum_{k=0}^{\infty} p_k z^k$. Then, using the method of supplementary variables as in Problem 5.7, it may be shown that

$$P(z) = (1 - \rho)\frac{(1 - z)B^*(\lambda - \lambda G(z))}{B^*(\lambda - \lambda G(z)) - z}$$

with

$$p_0 = 1 - \rho$$

(see [COHE 69] page 375, Equation (2.13) for details). Note that $P(z) \neq Q(z)$ and $p_0 \neq d_0$. The average number in system $\bar{N}$ (over all time) is simply

$$\bar{N} = P^{(1)}(1) = \rho + \frac{\rho^2(1 + C_b^2)}{2(1 - \rho)} + \frac{\rho}{2(1 - \rho)}[\bar{g}(1 + C_g^2) - 1]$$

The average queue size $\bar{N}_q$ is

$$\bar{N}_q = \sum_{k=1}^{\infty} (k - 1)p_k = \bar{N} - (1 - p_0) = \bar{N} - \rho$$

or

$$\overline{N}_q = \frac{\rho^2(1 + C_b^2)}{2(1 - \rho)} + \frac{\rho}{2(1 - \rho)}[\overline{g}(1 + C_g^2) - 1]$$

We may now apply Little's result to $\overline{N}_q$ to find the average wait W. (Note that Little's result *cannot* be applied to $\overline{q}$, the average number in system at *departure* instants.) So $\overline{N}_q = \overline{\lambda} W$, where $\overline{\lambda} = \lambda \overline{g}$. Thus the average wait W is

$$W = \frac{\rho \overline{x}(1 + C_b^2)}{2(1 - \rho)} + \overline{x}\frac{\overline{g}(1 + C_g^2) - 1}{2(1 - \rho)}$$

and so

$$\frac{W}{\overline{x}} = \frac{\rho(1 + C_b^2)}{2(1 - \rho)} + \frac{\overline{g}(1 + C_g^2) - 1}{2(1 - \rho)}$$

We know, from part (d), that

$$\frac{W_g}{\overline{x}} = \frac{\rho \overline{g}}{2(1 - \rho)}\left[1 + C_g^2 + \frac{C_b^2}{\overline{g}}\right]$$

Thus

$$\frac{W_s}{\overline{x}} = \frac{W - W_g}{\overline{x}} = \frac{\overline{g}(1 + C_g^2) - 1}{2(1 - \rho)} - \frac{\rho \overline{g}}{2(1 - \rho)}(1 + C_g^2) + \frac{\rho}{2(1 - \rho)}$$

or

$$\frac{W_s}{\overline{x}} = \frac{1}{2}\overline{g}(1 + C_g^2) - \frac{1}{2}$$

as before.]   □

PROBLEM 5.13

Consider an M/G/1 system in which service is instantaneous but is only available at "service instants," the intervals between successive service instants being independently distributed with PDF $F(x)$. The maximum number of customers that can be served at any service instant is m. Note that this is a bulk service system.

(a) Show that if q_n is the number of customers in the system just before the nth service instant, then

$$q_{n+1} = \begin{cases} q_n + v_n - m & q_n \geq m \\ v_n & q_n < m \end{cases}$$

where v_n is the number of arrivals in the interval between the nth and $(n + 1)$th service instants.

(b) Prove that the probability generating function of v_n is $F^*(\lambda - \lambda z)$. Hence show that $Q(z)$ is

$$Q(z) = \frac{\sum_{k=0}^{m-1} p_k(z^m - z^k)}{z^m[F^*(\lambda - \lambda z)]^{-1} - 1}$$

where $p_k = P[\bar{q} = k]$ $(k = 0, \cdots, m-1)$.

(c) The $\{p_k\}$ can be determined from the condition that within the unit disk of the z-plane, the numerator must vanish when the denominator does. Hence show that if $F(x) = 1 - e^{-\mu x}$,

$$Q(z) = \frac{z_m - 1}{z_m - z}$$

where z_m is the zero of $z^m[1 + \lambda(1-z)/\mu] - 1$ outside the unit disk.

SOLUTION

(a) If $q_n \geq m$, then m customers are served and clearly $q_{n+1} = q_n - m + v_n$. If $q_n < m$, everyone in the system gets served and $q_{n+1} = v_n$. Thus

$$q_{n+1} = \begin{cases} q_n + v_n - m & q_n \geq m \\ v_n & q_n < m \end{cases}$$

For convenience, we define the function $\Delta_{m,k}$ by

$$\Delta_{m,k} \overset{\Delta}{=} \begin{cases} m & k \geq m \\ k & 0 \leq k \leq m \end{cases}$$

Thus

$$q_{n+1} = q_n - \Delta_{m,q_n} + v_n$$

(b) Since $F(x)$ is the distribution between successive service instants and $V(z)$ is the z-transform for the (Poisson) arrival process, the argument at the end of Problem 5.11 may be applied. Thus $V(z) = F^*(\lambda - \lambda z)$. To find $Q(z)$, we proceed as follows.

$$E\left[z^{q_{n+1}}\right] = E\left[z^{q_n - \Delta_{m,q_n} + v_n}\right] = E\left[z^{q_n - \Delta_{m,q_n}}\right]E\left[z^{v_n}\right]$$

since v_n, q_n are independent. Let $n \to \infty$ and obtain

$$Q(z) = E\left[z^{\bar{q}}\right] = E\left[z^{\bar{q} - \Delta_{m,\bar{q}}}\right]E\left[z^{\bar{v}}\right]$$
$$Q(z) = V(z)E\left[z^{\bar{q} - \Delta_{m,\bar{q}}}\right]$$

But

$$E\left[z^{\tilde{q}\cdot\Delta_{m,\tilde{q}}}\right] = \sum_{k=0}^{m-1} P[\tilde{q} = k]z^{k-k} + \sum_{k=m}^{\infty} P[\tilde{q} = k]z^{k-m}$$

$$= \sum_{k=0}^{m-1} p_k + \frac{1}{z^m} \sum_{k=m}^{\infty} p_k z^k$$

$$= \sum_{k=0}^{m-1} p_k + \frac{1}{z^m} \left(Q(z) - \sum_{k=0}^{m-1} p_k z^k \right)$$

Thus

$$Q(z) = V(z) \left[\frac{1}{z^m} Q(z) + \sum_{k=0}^{m-1} p_k \left(1 - \frac{z^k}{z^m} \right) \right]$$

Recalling that $V(z) = F^*(\lambda - \lambda z)$ we have

$$Q(z) = F^*(\lambda - \lambda z) \frac{\sum_{k=0}^{m-1} p_k(z^m - z^k)}{z^m - F^*(\lambda - \lambda z)}$$

or

$$Q(z) = \frac{\sum_{k=0}^{m-1} p_k(z^m - z^k)}{z^m[F^*(\lambda - \lambda z)]^{-1} - 1}$$

(c) Assume that $F(x) = 1 - e^{-\mu x}$. Therefore $F^*(\lambda - \lambda z) = \mu/(\lambda - \lambda z + \mu)$ and

$$Q(z) = \frac{\sum_{k=0}^{m-1} p_k(z^m - z^k)}{z^m \left[\dfrac{\lambda - \lambda z + \mu}{\mu} \right] - 1}$$

$$Q(z) = \frac{\sum_{k=0}^{m-1} p_k(z^m - z^k)}{z^m \left[1 + \dfrac{\lambda(1 - z)}{\mu} \right] - 1} \triangleq \frac{P(z)}{D(z)}$$

We now examine the roots of the denominator

$$D(z) = z^m \left[1 + \frac{\lambda(1 - z)}{\mu} \right] - 1$$

Rewrite $D(z)$ as

$$D(z) = - \left[\frac{\lambda}{\mu} z^{m+1} - \left(1 + \frac{\lambda}{\mu} \right) z^m + 1 \right]$$

$$= -[m\rho z^{m+1} - (1 + m\rho)z^m + 1]$$

where we write $\rho = \lambda/m\mu$. For stability we require $\lambda < m\mu$, or $\rho < 1$. We note that $D(z)$ is the negative of the polynomial in Eq. (1.84) (with m replacing r), and so we may use the results of Problem 4.10. Thus $D(z)$ has one root at $z = 1$, $m - 1$ roots in the range $|z| < 1$, and one root (say, z_m) in $|z| > 1$. By analyticity of $Q(z)$ for $|z| \le 1$, the m roots of $D(z)$ satisfying $|z| \le 1$ cancel with the m numerator roots leaving

$$Q(z) = K \frac{1}{z_m - z}$$

Now $Q(1) = 1$ implies $K = z_m - 1$. Thus

$$Q(z) = \frac{z_m - 1}{z_m - z}$$ □

PROBLEM 5.14

Consider an M/G/1 system with bulk service. Whenever the server becomes free, he accepts *two* customers from the queue into service simultaneously, or, if only one is on queue, he accepts that one; in either case, the service time for the group (of size 1 or 2) is taken from $B(x)$. Let q_n be the number of customers remaining after the nth service instant. Let v_n be the number of arrivals during the nth service. Define $B^*(s)$, $Q(z)$, and $V(z)$ as transforms associated with the random variables $\tilde{x}$, $\tilde{q}$, and $\tilde{v}$ as usual. Let $\rho = \lambda \bar{x}/2$.

(a) Using the method of imbedded Markov chains, find

$$E(\tilde{q}) = \lim_{n \to \infty} E(q_n)$$

in terms of ρ, $\sigma_b{}^2$, and $P[\tilde{q} = 0] \overset{\Delta}{=} p_0$.

(b) Find $Q(z)$ in terms of $B^*(\cdot)$, p_0, and $p_1 \overset{\Delta}{=} P[\tilde{q} = 1]$.

(c) Express p_1 in terms of p_0.

SOLUTION

(a) Clearly we may write

$$q_{n+1} = \begin{cases} q_n - 2 + v_{n+1} & q_n \ge 2 \\ q_n - 1 + v_{n+1} & q_n = 1 \\ v_{n+1} & q_n = 0 \end{cases}$$

Introducing the function

$$\Delta_{2,k} \overset{\Delta}{=} \begin{cases} 2 & k \ge 2 \\ k & 0 \le k \le 2 \end{cases}$$

we have $q_{n+1} = q_n - \Delta_{2.q_n} + v_{n+1}$. Letting $n \to \infty$ and taking expectations

$$\bar{q} = \bar{q} - E\left[\Delta_{2.\tilde{q}}\right] + \bar{v}$$

But

$$E\left[\Delta_{2.\tilde{q}}\right] = \sum_{k=0}^{\infty} \Delta_{2.k} P[\tilde{q} = k]$$

$$= P[\tilde{q} = 1] + \sum_{k=2}^{\infty} 2P[\tilde{q} = k]$$

$$= p_1 + 2(1 - p_0 - p_1)$$

So

$$\bar{v} = E\left[\Delta_{2.\tilde{q}}\right] = 2 - 2p_0 - p_1$$

Recall that

$$q_{n+1} = q_n - \Delta_{2.q_n} + v_{n+1}$$

Squaring this equation gives

$$q_{n+1}^2 = q_n^2 - 2q_n\Delta_{2.q_n} + \Delta_{2.q_n}^2 + 2v_{n+1}(q_n - \Delta_{2.q_n}) + v_{n+1}^2$$

Let $n \to \infty$ and take expectations

$$\overline{q^2} = \overline{q^2} - 2E\left[\tilde{q}\Delta_{2.\tilde{q}}\right] + E\left[\Delta_{2.\tilde{q}}^2\right] + 2\bar{v}E\left[\tilde{q} - \Delta_{2.\tilde{q}}\right] + \overline{v^2}$$

since $\tilde{v}$ and $\tilde{q}$ are independent. So

$$2E\left[\tilde{q}\Delta_{2.\tilde{q}}\right] = E\left[\Delta_{2.\tilde{q}}^2\right] + \overline{v^2} + 2\bar{v}\left(\bar{q} - E\left[\Delta_{2.\tilde{q}}\right]\right)$$

Now $E\left[\Delta_{2.\tilde{q}}\right] = \bar{v}$ and

$$E\left[\Delta_{2.\tilde{q}}^2\right] = \sum_{k=1}^{\infty} \Delta_{2.k}^2 P[\tilde{q} = k]$$

$$= P[\tilde{q} = 1] + 4\sum_{k=2}^{\infty} P[\tilde{q} = k]$$

$$= p_1 + 4(1 - p_0 - p_1)$$

Also

$$E\left[\tilde{q}\Delta_{2,\tilde{q}}\right] = \sum_{k=1}^{\infty} k\Delta_{2,k}P[\tilde{q} = k]$$

$$= P[\tilde{q} = 1] + \sum_{k=2}^{\infty} 2kP[\tilde{q} = k]$$

$$= p_1 + 2\sum_{k=1}^{\infty} kP[\tilde{q} = k] - 2p_1$$

$$= 2\bar{q} - p_1$$

Thus

$$2(2\bar{q} - p_1) = 4 - 4p_0 - 3p_1 + \overline{v^2} + 2\bar{v}(\bar{q} - \bar{v})$$

Therefore, using $p_1 = 2 - 2p_0 - \bar{v}$, we find

$$\bar{q} = \frac{2 - 2p_0 + \bar{v} + \overline{v^2} - 2(\bar{v})^2}{4 - 2\bar{v}}$$

From Problem 5.11, we know that $V(z) = B^*(\lambda - \lambda z)$. Then, by differentiation, we have

$$\bar{v} = V^{(1)}(1) = \lambda\bar{x} = 2\rho$$

and

$$\overline{v^2} - \bar{v} = V^{(2)}(1) = \lambda^2\overline{x^2}$$

Thus

$$\bar{q} = \frac{2(1 - p_0) + 2\rho + \lambda^2\overline{x^2} + 2\rho - 2(4\rho^2)}{4 - 4\rho}$$

$$\bar{q} = \rho + \frac{2(1 - p_0) + \lambda^2\overline{x^2} - 4\rho^2}{4(1 - \rho)}$$

$$\bar{q} = \rho + \frac{2(1 - p_0) + \lambda^2\sigma_b^2}{4(1 - \rho)}$$

(b) We have

$$Q(z) = E\left[z^{\tilde{q}}\right] = E\left[z^{\tilde{q} - \Delta_{2,\tilde{q}} + \tilde{v}}\right]$$

$$= E\left[z^{\tilde{v}}\right]E\left[z^{\tilde{q} - \Delta_{2,\tilde{q}}}\right]$$

$$= V(z)E\left[z^{\tilde{q} - \Delta_{2,\tilde{q}}}\right]$$

But

$$E\left[z^{\tilde{q} \ \Delta_2 \tilde{q}}\right] = \sum_{k=0}^{\infty} z^{k-\Delta_{2k}} P[\tilde{q} = k]$$

$$= P[\tilde{q} = 0] + P[\tilde{q} = 1] + \sum_{k=2}^{\infty} z^{k-2} P[\tilde{q} = k]$$

$$= p_0 + p_1 + \frac{1}{z^2}[Q(z) - p_0 - p_1 z]$$

Thus

$$Q(z) = V(z)\left[p_0 + p_1 + \frac{1}{z^2}[Q(z) - p_0 - p_1 z]\right]$$

$$Q(z) = V(z)\frac{p_0(1 - z^2) + p_1 z(1 - z)}{V(z) - z^2}$$

Finally

$$Q(z) = B^*(\lambda - \lambda z)\frac{p_0(1 - z^2) + p_1 z(1 - z)}{B^*(\lambda - \lambda z) - z^2}$$

(c) From part (a),

$$\bar{v} = 2 - 2p_0 - p_1$$

But, from part (b),

$$\bar{v} = \lambda\bar{x} = 2\rho$$

Equating these two expressions gives

$$p_1 = 2(1 - p_0 - \rho)$$

or

$$p_1 = 2(1 - p_0) - \lambda\bar{x} \qquad\qquad \square$$

PROBLEM 5.15

Consider an M/G/1 queueing system with the following variation. The server refuses to serve any customers unless at least two customers are ready for service, at which time both are "taken into" service. These two customers are served individually and independently, one after the other. The instant at which the second of these two is finished is called a "critical" time and we shall use these critical times as the points in an imbedded Markov chain. Immediately following a critical time, if there are two more ready for service, they are both "taken into" service as above. If one or none is ready, then the server waits until a pair is ready, and so on. Let

q_n = number of customers left behind in the system immediately following the nth critical time

v_n = number of customers arriving during the combined service time of the nth *pair* of customers

(a) Derive a relationship between q_{n+1}, q_n, and v_{n+1}.

(b) Find

$$V(z) = \sum_{k=0}^{\infty} P[v_n = k]z^k$$

(c) Derive an expression for $Q(z) = \lim_{n \to \infty} Q_n(z)$ in terms of $p_0 = P[\tilde{q} = 0]$, where

$$Q_n(z) = \sum_{k=0}^{\infty} P[q_n = k]z^k$$

(d) How would you solve for p_0?

(e) Describe (do *not* calculate) two methods for finding $\bar{q}$.

SOLUTION

(a) Since the server refuses to serve any customers unless at least two are in the system, we see that

$$q_{n+1} = \begin{cases} q_n - 2 + v_{n+1} & q_n \geq 2 \\ v_{n+1} & q_n \leq 1 \end{cases}$$

Introducing the function

$$\Delta_{2,k} \triangleq \begin{cases} 2 & k \geq 2 \\ k & 0 \leq k \leq 2 \end{cases}$$

we have

$$q_{n+1} = q_n - \Delta_{2,q_n} + v_{n+1}$$

(b) $V(z)$ is the z-transform for the number of arrivals in an interval, which is the sum of two service times. The transform of the pdf for this interval is clearly $[B^*(s)]^2$. By analogy to the development in Problem 5.11, we see that

$$V(z) = [B^*(\lambda - \lambda z)]^2$$

(c) From part (a) we have

$$Q_{n+1}(z) = E\left[z^{q_{n+1}}\right] = E\left[z^{q_n - \Delta_{2,q_n} + v_{n+1}}\right] = E\left[z^{v_{n+1}}\right]E\left[z^{q_n - \Delta_{2,q_n}}\right]$$

by independence of q_n and v_{n+1}. Letting $n \to \infty$, we have

$$Q(z) = E\left[z^{\tilde{q}}\right] = V(z)E\left[z^{\tilde{q} - \Delta_{2,\tilde{q}}}\right]$$

But

$$E\left[z^{\tilde{q} - \Delta_{2,\tilde{q}}}\right] = \sum_{k=0}^{\infty} z^{k - \Delta_{2,k}} P[\tilde{q} = k]$$

$$= p_0 + p_1 + \sum_{k=2}^{\infty} z^{k-2} p_k$$

$$= p_0 + p_1 + \frac{1}{z^2}[Q(z) - p_0 - p_1 z]$$

Thus

$$Q(z) = \frac{V(z)}{z^2}\left[Q(z) + (z^2 - 1)p_0 + (z^2 - z)p_1\right]$$

$$Q(z) = V(z)\frac{(1 - z^2)p_0 + (z - z^2)p_1}{V(z) - z^2}$$

To eliminate p_1 we proceed as follows:

$$1 = \frac{Q(1)}{V(1)} = \lim_{z \to 1} \frac{(1 - z^2)p_0 + (z - z^2)p_1}{V(z) - z^2}$$

Using L'Hospital's rule and $V^{(1)}(1) = \bar{v}$, we find

$$\bar{v} = 2 - 2p_0 - p_1$$

(This could also be obtained from the equation $\bar{q} = \bar{q} + E\left[\Delta_{2,\tilde{q}}\right] + \bar{v}$.) But $V(z) = [B^*(\lambda - \lambda z)]^2$, so that

$$\bar{v} = V^{(1)}(1) = 2\lambda\bar{x}$$

Thus

$$2\lambda\bar{x} = 2 - 2p_0 - p_1$$

and

$$p_1 = 2 - 2p_0 - 2\lambda\bar{x}$$

Substituting this expression for p_1 into $Q(z)$, we have

$$Q(z) = [B^*(\lambda - \lambda z)]^2 \frac{(1 - z)[p_0(1 - z) + 2z(1 - \lambda\bar{x})]}{[B^*(\lambda - \lambda z)]^2 - z^2}$$

(d) Equate roots of the denominator of $Q(z)$ with that of the numerator for $|z| < 1$ using analyticity of $Q(z)$.

(e) (1) Use the relation $\bar{q} = Q^{(1)}(1)$.

(2) Square the equation in part (a), let $n \to \infty$, and take expectations.   □

PROBLEM 5.16

Consider an M/G/1 queueing system in which service is given as follows. Upon entry into service, a coin is tossed, which has probability p of giving Heads. If the result is Heads, then the service time for that customer is zero seconds. If Tails, his service time is drawn from the following exponential distribution:

$$pe^{-px} \qquad x \geq 0$$

(a) Find the average service time $\bar{x}$.

(b) Find the variance of service time σ_b^2.

(c) Find the expected waiting time W.

(d) Find $W^*(s)$.

(e) From (d), find the expected waiting time W.

(f) From (d), find $W(t) = P[\text{waiting time} \leq t]$.

SOLUTION

The service time density $b(x)$ is given by

$$b(x) = pu_0(x) + (1 - p)pe^{-px} \qquad x \geq 0$$

(a) The mean service time is

$$\bar{x} = 0 \cdot p + \frac{1}{p} \cdot (1 - p) = \frac{1 - p}{p}$$

[Thus, for stability, we require $p > \lambda/(\lambda + 1)$.]

(b) The second moment of service time is

$$\overline{x^2} = 0 \cdot p + \frac{2}{p^2} \cdot (1 - p) = \frac{2(1 - p)}{p^2}$$

Thus

$$\sigma_b^2 = \overline{x^2} - (\bar{x})^2 = \frac{2(1 - p)}{p^2} - \frac{(1 - p)^2}{p^2}$$

$$\sigma_b^2 = \frac{1 - p^2}{p^2}$$

(c) For M/G/1, $W = \lambda \overline{x^2}/2(1 - p)$ by Eq. (1.101). Using $\rho = \lambda \bar{x} = \lambda(1 - p)/p$, we have

$$W = \frac{\lambda \dfrac{2(1 - p)}{p^2}}{2\left(1 - \dfrac{\lambda(1 - p)}{p}\right)}$$

$$W = \frac{\rho}{p(1 - \rho)}$$

(d) We first note that

$$B^*(s) = p + (1 - p)\frac{p}{s + p} = \frac{p(s + 1)}{s + p}$$

Thus, by Eq. (1.105),

$$W^*(s) = \frac{s(1 - \rho)}{s - \lambda + \lambda \dfrac{p(s + 1)}{s + p}}$$

$$W^*(s) = \frac{(1 - \rho)(s + p)}{s + p(1 - \rho)}$$

(e) Differentiating gives

$$W^{*(1)}(s) = (1 - \rho)\left[\frac{[s + p(1 - \rho)] - (s + p)}{[s + p(1 - \rho)]^2}\right]$$

$$= (1 - \rho)\frac{-p\rho}{[s + p(1 - \rho)]^2}$$

Thus

$$W = -W^{*(1)}(0) = \frac{\rho}{p(1 - \rho)}$$

[same as part (c)].

(f) We have

$$W^*(s) = \frac{(1 - \rho)(s + p)}{s + p(1 - \rho)}$$

$$= (1 - \rho) + \frac{p\rho(1 - \rho)}{s + p(1 - \rho)}$$

Inverting we find the pdf as

$$w(y) = (1 - \rho)u_0(y) + p\rho(1 - \rho)e^{-p(1 - \rho)y} \qquad y \geq 0$$

Thus the PDF is

$$W(y) = 1 - \rho e^{-p(1 - \rho)y} \qquad y \geq 0 \qquad\qquad \square$$

PROBLEM 5.17

Consider an M/G/1 queue. Let E be the event that T sec have elapsed since the arrival of the last customer. Begin at an arrival time and measure the time $\tilde{w}$ until event E next occurs. This measurement may involve the observation of many customer arrivals before E occurs.

  (a) Let $\hat{A}(t)$ be the interarrival-time distribution for those intervals during which E does *not* occur. Find $\hat{A}(t)$.
  (b) Find $\hat{A}^*(s) = \int_0^\infty e^{-st}\, d\hat{A}(t)$.
  (c) Find $W^*(s \mid n) = \int_0^\infty e^{-sw}\, dW(w \mid n)$, where $W(w \mid n) = P[\text{time to event } E \le w \mid n \text{ arrivals occur before } E]$.
  (d) Find $W^*(s) = \int_0^\infty e^{-sw}\, dW(w)$, where $W(w) = P[\text{time to event } E \le w]$.
  (e) Find the mean time to event E.

SOLUTION

  (a) If E does not occur, then the interarrival time must be $< T$ seconds. The probability that an interarrival time is $< T$ seconds is $1 - e^{-\lambda T}$. Thus

$$\hat{A}(t) = \begin{cases} \dfrac{1 - e^{-\lambda t}}{1 - e^{-\lambda T}} & 0 \le t < T \\ 1 & t \ge T \end{cases}$$

$$\hat{a}(t) = \begin{cases} \dfrac{\lambda e^{-\lambda t}}{1 - e^{-\lambda T}} & 0 \le t < T \\ 0 & t \ge T \end{cases}$$

  (b) We have

$$\hat{A}^*(s) = \int_0^\infty e^{-st}\hat{a}(t)\, dt = \int_0^T e^{-st}\frac{\lambda e^{-\lambda t}}{1 - e^{-\lambda T}}\, dt$$

$$\hat{A}^*(s) = \frac{\lambda}{s + \lambda} \cdot \frac{1 - e^{-(s+\lambda)T}}{1 - e^{-\lambda T}}$$

  (c) Recall that the random variable $\tilde{w}$ is defined to be the time to event E.

$$W(w \mid n) = P[\tilde{w} \le w \mid n \text{ arrivals occur before } E]$$

Since exactly n arrivals occur before E, we know that the sum of these n interarrival times plus T must be the value of $\tilde{w}$. Thus

$$W(w \mid n) = P[n \text{ interarrival times} + T \le w]$$

$$= P[t_1 + \cdots + t_n + T \le w]$$

Therefore

$$W^*(s \mid n) = E\left[e^{-s(t_1 + \cdots + t_n + T)}\right]$$

Noting that each interarrival time t_i occurs before E and thus has transform $\hat{A}^*(s)$, we have

$$W^*(s \mid n) = e^{-sT}[\hat{A}^*(s)]^n$$

(d) Unconditioning gives

$$W^*(s) = \sum_{n=0}^{\infty} W^*(s \mid n)P[n \text{ arrivals occur before } E]$$

$$= \sum_{n=0}^{\infty} e^{-sT}[\hat{A}^*(s)]^n[1 - e^{-\lambda T}]^n e^{-\lambda T}$$

$$= e^{-(s+\lambda)T} \frac{1}{1 - \hat{A}^*(s)[1 - e^{-\lambda T}]}$$

$$= e^{-(s+\lambda)T} \frac{1}{1 - \dfrac{\lambda}{s+\lambda}[1 - e^{-(s+\lambda)T}]}$$

Thus

$$W^*(s) = e^{-(s+\lambda)T} \frac{s + \lambda}{s + \lambda e^{-(s+\lambda)T}}$$

(e) We may calculate the mean time W to event E by $W = -W^{*(1)}(0)$. Instead let us find W by recognizing that $W = \bar{N}\bar{\hat{t}} + T$, where $\bar{\hat{t}}$ is the mean interarrival time from the distribution $\hat{A}(t)$ and $\bar{N}$ is the mean number of such arrivals before E. But

$$\bar{N} = \sum_{k=1}^{\infty} k[1 - e^{-\lambda T}]^k e^{-\lambda T} = e^{-\lambda T}[1 - e^{-\lambda T}]\frac{1}{[e^{-\lambda T}]^2}$$

$$\bar{N} = e^{\lambda T}[1 - e^{-\lambda T}]$$

$$\bar{\hat{t}} = \int_0^{\infty} [1 - \hat{A}(t)] \, dt = \int_0^T \frac{e^{-\lambda t} - e^{-\lambda T}}{1 - e^{-\lambda T}} \, dt$$

$$\bar{\hat{t}} = \frac{1}{1 - e^{-\lambda T}} \left(\frac{1 - e^{-\lambda T}}{\lambda} - Te^{-\lambda T} \right)$$

Thus

$$W = e^{\lambda T}[1 - e^{-\lambda T}]\frac{1}{1 - e^{-\lambda T}} \left(\frac{1 - e^{-\lambda T}}{\lambda} - Te^{-\lambda T} \right) + T$$

$$= \frac{e^{\lambda T} - 1}{\lambda} - T + T$$

$$W = \frac{e^{\lambda T} - 1}{\lambda} \qquad \qquad \square$$

PROBLEM 5.18

Consider an M/G/1 system in which time is divided into intervals of length q sec each. Assume that *arrivals* are Bernoulli, that is,

$$P[1 \text{ arrival in any interval}] = \lambda q$$

$$P[0 \text{ arrivals in any interval}] = 1 - \lambda q$$

$$P[> 1 \text{ arrival in any interval}] = 0$$

Assume that a customer's *service time* $\tilde{x}$ is some multiple of q sec such that

$$P[\text{service time} = nq \text{ sec}] = g_n \qquad n = 0, 1, 2, \ldots$$

(a) Find $E[\text{number of arrivals in an interval}]$.
(b) Find the average arrival rate.
(c) Express $E[\tilde{x}] \stackrel{\Delta}{=} \bar{x}$ and $E[\tilde{x}(\tilde{x} - q)] \stackrel{\Delta}{=} \overline{x^2} - \overline{x}q$ in terms of the moments of the g_n distribution (i.e., let $\overline{g^k} \stackrel{\Delta}{=} \sum_{n=0}^{\infty} n^k g_n$).
(d) Find $y_{mn} = P[m \text{ customers arrive in } nq \text{ sec}]$.
(e) Let $v_m = P[m \text{ customers arrive during the service of a customer}]$ and let

$$V(z) = \sum_{m=0}^{\infty} v_m z^m \qquad \text{and} \qquad G(z) = \sum_{m=0}^{\infty} g_m z^m$$

Express $V(z)$ in terms of $G(z)$ and the system parameters λ and q.
(f) Find the mean number of arrivals during a customer service time from (e).

SOLUTION

(a) We calculate

$$E[\text{number of arrivals in an interval}] = 1 \cdot \lambda q + 0 \cdot (1 - \lambda q) + 0 = \lambda q$$

(b) We have

$$\text{average arrival rate} = \frac{\lambda q}{q} = \lambda$$

(c) The moments are

$$\bar{x} = \sum_{n=0}^{\infty} (nq)g_n = q\bar{g}$$

$$\overline{x^2} = \sum_{n=0}^{\infty} (nq)^2 g_n = q^2 \overline{g^2}$$

Thus

$$\overline{x^2} - \bar{x}q = q^2(\overline{g^2} - \bar{g})$$

(d) In nq seconds there are n intervals. We want m intervals each with one arrival and $n - m$ with none. Thus

$$y_{mn} = \binom{n}{m}(\lambda q)^m (1 - \lambda q)^{n-m} \qquad m \le n$$

$$(y_{mn} = 0 \qquad \text{for } m > n)$$

(e) Condition on the service time $\tilde{x}$. Then

$$v_m = \sum_{n=0}^{\infty} g_n P[v_m \mid \tilde{x} = nq] = \sum_{n=0}^{\infty} g_n y_{mn}$$

and thus

$$V(z) = \sum_{m=0}^{\infty} v_m z^m = \sum_{m=0}^{\infty} \sum_{n=0}^{\infty} g_n y_{mn} z^m$$

$$= \sum_{m=0}^{\infty} \sum_{n=m}^{\infty} g_n z^m \binom{n}{m}(\lambda q)^m (1 - \lambda q)^{n-m}$$

$$= \sum_{n=0}^{\infty} g_n \sum_{m=0}^{n} \binom{n}{m}(\lambda q z)^m (1 - \lambda q)^{n-m}$$

$$= \sum_{n=0}^{\infty} g_n (\lambda q z + 1 - \lambda q)^n$$

$$V(z) = G(1 - \lambda q(1 - z))$$

(f) Differentiating gives

$$\bar{v} = V^{(1)}(1) = G^{(1)}(1 - \lambda q(1 - z))(\lambda q)\big|_{z=1}$$

$$= G^{(1)}(1)(\lambda q)$$

$$\bar{v} = \lambda q \bar{g}$$

That is,

$$\bar{v} = \lambda \bar{x} = \rho \qquad \qquad \Box$$

PROBLEM 5.19

Suppose that in an M/G/1 queueing system the *cost* of making a customer wait t sec is $c(t)$ dollars, where $c(t) = \alpha e^{\beta t}$. Find the average cost of queueing for a customer. Also determine the conditions necessary to keep the average cost finite.

SOLUTION

Clearly β is constrained to be real. The average cost of queueing is simply

$$\bar{c} = \int_0^\infty c(t)\,dW(t)$$

$$\bar{c} = \alpha \int_0^\infty e^{\beta t}\,dW(t)$$

Assuming this integral exists, $\bar{c} = \alpha W^*(-\beta)$. For M/G/1, we may use the P-K transform equation [Eq. (1.105)] to obtain

$$\bar{c} = \alpha \frac{(-\beta)(1-\rho)}{-\beta - \lambda + \lambda B^*(-\beta)}$$

or

$$\bar{c} = \frac{\alpha\beta(1-\rho)}{\beta + \lambda - \lambda B^*(-\beta)}$$

We must next determine conditions for the existence of the integral

$$J \overset{\Delta}{=} \int_0^\infty e^{\beta t}\,dW(t)$$

Using the definition of the Laplace transform, we show below that there is a value β_0 such that $W^*(-\beta) < \infty$ for $\beta < \beta_0$. We seek the maximum such β_0. Since J clearly exists for $\beta \leq 0$, we need merely study $0 < \beta < \beta_0$. If $J < \infty$, then for M/G/1 we have

$$J = W^*(-\beta) = \frac{\beta(1-\rho)}{\beta + \lambda - \lambda B^*(-\beta)}$$

In this case, β_0 must then be the smallest positive pole of $W^*(-\beta)$. That is, β_0 is the smallest positive root of the equation $\beta + \lambda - \lambda B^*(-\beta) = 0$, and so it must be that

$$\beta_0 = \lambda[B^*(-\beta_0) - 1]$$

Let us now give a rigorous argument for the existence of such a β_0. Define

$$f(\beta) \overset{\Delta}{=} \lambda[B^*(-\beta) - 1]$$

and note (by differentiation) that $f(\beta)$ is convex increasing in β, $f(0) = 0$, and $f'(0) = \rho < 1$ for a stable system. Also,

$$f'(\beta) = \lambda \left(\int_0^\infty x e^{\beta x} b(x)\,dx \right)$$

$$= \lambda \left(\int_0^\infty x \left[\sum_{k=0}^\infty \frac{(\beta x)^k}{k!} \right] b(x)\,dx \right)$$

$$= \lambda \sum_{k=0}^\infty \frac{\overline{x^{k+1}}}{k!} \beta^k$$

Thus, for $\beta \geq 0$,

$$f'(\beta) \geq \lambda\left[\bar{x} + \beta\overline{x^2}\right] \geq \lambda\left[\bar{x} + \beta(\bar{x})^2\right]$$

So, for $\lambda > 0$ and $\bar{x} > 0$ (i.e., $\rho > 0$), we have

$$\lim_{\beta \to \infty} f'(\beta) = \infty$$

Thus, as shown in the following figure, the equation $\beta = f(\beta)$ has a solution $\beta_0 > 0$ for $0 < \rho < 1$.

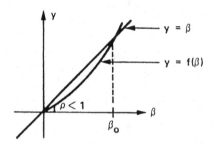

In summary, for M/G/1 ($0 < \rho < 1$)

$$\bar{c} = \frac{\alpha\beta(1 - \rho)}{\beta + \lambda - \lambda B^*(-\beta)}$$

and $\bar{c}$ is finite for $\beta < \beta_0$, where β_0 is the smallest positive root of the equation $\beta + \lambda - \lambda B^*(-\beta) = 0$. □

PROBLEM 5.20

We wish to find the *interdeparture* time probability density function $d(t)$ for an M/G/1 queueing system.

(a) Find the Laplace transform $D^*(s)$ of this density conditioned first on a nonempty queue left behind, and second on an empty queue left behind by a departing customer. Combine these results to get the Laplace transform of the interdeparture time density and from this find the density itself.

(b) Give an explicit form for the probability distribution $D(t)$, or density $d(t) = dD(t)/dt$, of the interdeparture time when we have a constant service time, that is,

$$B(x) = \begin{cases} 0 & x < T \\ 1 & x \geq T \end{cases}$$

SOLUTION

(a) By conditioning on whether or not the departure leaves behind an empty queue, we have

$$D^*(s)\big|_{\substack{\text{nonempty queue} \\ \text{left behind}}} = B^*(s)$$

$$D^*(s)\big|_{\substack{\text{empty queue} \\ \text{left behind}}} = B^*(s)A^*(s) = B^*(s)\frac{\lambda}{s+\lambda}$$

Thus, unconditioning, we have

$$D^*(s) = \rho B^*(s) + (1-\rho)B^*(s)\frac{\lambda}{s+\lambda}$$

or

$$D^*(s) = \frac{\rho s + \lambda}{s+\lambda} B^*(s)$$

From the first of these two forms we invert by inspection to obtain the interdeparture time density as

$$d(t) = \rho b(t) + (1-\rho)b(t) \circledast \lambda e^{-\lambda t} \qquad t \geq 0$$

where $\circledast$ represents convolution.

(b) Since we have a constant service time of T seconds, we have

$$b(t) = u_0(t-T) \Leftrightarrow e^{-sT} = B^*(s)$$

We may proceed from the expression for $D^*(s)$ or from $d(t)$ in part (a). Let us use the latter approach to find

$$b(t) \circledast \lambda e^{-\lambda t} = \int_0^t b(t-x)\lambda e^{-\lambda x}\,dx$$

$$= \int_0^t u_0(t-x-T)\lambda e^{-\lambda x}\,dx$$

$$= \lambda e^{-\lambda(t-T)}\delta(t-T)$$

So

$$d(t) = \rho u_0(t-T) + (1-\rho)\lambda e^{-\lambda(t-T)}\delta(t-T)$$

and thus

$$D(t) = \begin{cases} 0 & t < T \\ 1 - (1-\rho)e^{-\lambda(t-T)} & t \geq T \end{cases} \qquad \square$$

PROBLEM 5.21

Consider an M/G/1 system in which a departing customer immediately joins the queue again with probability p, or departs forever with probability $q = 1 - p$. Service is FCFS, and the service time for a returning customer is independent of his previous service times. Let $B^*(s)$ be the transform for the service time pdf and let $B_T^*(s)$ be the transform for a customer's *total* service time pdf.

(a) Find $B_T^*(s)$ in terms of $B^*(s)$, p, and q.

(b) Let $\overline{x_T^n}$ be the nth moment of the total service time. Find $\overline{x_T^1}$ and $\overline{x_T^2}$ in terms of $\overline{x}$, $\overline{x^2}$, p, and q.

(c) Show that the following recurrence formula holds:

$$\overline{x_T^n} = \overline{x^n} + \frac{p}{q} \sum_{k=1}^{n} \binom{n}{k} \overline{x^k}\, \overline{x_T^{n-k}}$$

(d) Let

$$Q_T(z) = \sum_{k=0}^{\infty} p_{kT} z^k$$

where $p_{kT} = P[\text{number in system} = k]$. For $\lambda \overline{x} < q$ prove that

$$Q_T(z) = \left(1 - \frac{\lambda \overline{x}}{q}\right) \frac{q(1 - z)B^*(\lambda - \lambda z)}{(q + pz)B^*(\lambda - \lambda z) - z}$$

(e) Find $\overline{N}$, the average number of customers in the system.

SOLUTION

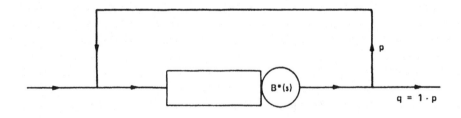

(a) Conditioning on the number of return trips, we have

$$B_T^*(s \mid \text{exactly } n \text{ return trips}) = [B^*(s)]^{n+1}$$

Unconditioning yields

$$B_T^*(s) = \sum_{n=0}^{\infty} B_T^*(s \mid \text{exactly } n \text{ return trips}) \cdot P(\text{exactly } n \text{ return trips})$$

$$= \sum_{n=0}^{\infty} [B^*(s)]^{n+1} q p^n = q B^*(s) \sum_{n=0}^{\infty} [p B^*(s)]^n$$

$$B_T^*(s) = \frac{q B^*(s)}{1 - p B^*(s)}$$

(b) Differentiating, we obtain

$$B_T^{*(1)}(s) = \frac{[1 - p B^*(s)] q B^{*(1)}(s) - q B^*(s)[-p B^{*(1)}(s)]}{[1 - p B^*(s)]^2}$$

$$= \frac{q B^{*(1)}(s)}{[1 - p B^*(s)]^2}$$

Thus

$$\overline{x_T^1} = -B_T^{*(1)}(0) = -\frac{q(-\overline{x})}{(1 - p)^2}$$

$$\overline{x_T^1} = \frac{\overline{x}}{q}$$

The second derivative gives

$$B_T^{*(2)}(s) = \frac{[1 - p B^*(s)]^2 q B^{*(2)}(s) - q B^{*(1)}(s) 2 [1 - p B^*(s)][-p B^{*(1)}(s)]}{[1 - p B^*(s)]^4}$$

$$= \frac{[1 - p B^*(s)] q B^{*(2)}(s) + 2 p q [B^{*(1)}(s)]^2}{[1 - p B^*(s)]^3}$$

Thus

$$\overline{x_T^2} = B_T^{*(2)}(0) = \frac{(1 - p) q \overline{x^2} + 2 p q (-\overline{x})^2}{(1 - p)^3}$$

$$\overline{x_T^2} = \frac{\overline{x^2}}{q} + \frac{2 p (\overline{x})^2}{q^2}$$

(c) From

$$B_T^*(s) = \frac{q B^*(s)}{1 - p B^*(s)}$$

we have

$$B_T^*(s) = q B^*(s) + p B^*(s) B_T^*(s)$$

Thus

$$B_T^{*(n)}(s) = qB^{*(n)}(s) + p[B^*(s)B_T^*(s)]^{(n)}$$

Using the formula for the nth derivative of the product of two functions, namely,

$$(fg)^{(n)} = \sum_{k=0}^{n} \binom{n}{k} f^{(k)} g^{(n-k)}$$

we have

$$B_T^{*(n)}(s) = qB^{*(n)}(s) + p \sum_{k=0}^{n} \binom{n}{k} B^{*(k)}(s) B_T^{*(n-k)}(s)$$

Setting $s = 0$ we obtain

$$B_T^{*(n)}(0) = qB^{*(n)}(0) + p \sum_{k=0}^{n} \binom{n}{k} B^{*(k)}(0) B_T^{*(n-k)}(0)$$

or

$$(-1)^n \overline{x_T^n} = q(-1)^n \overline{x^n} + p \sum_{k=0}^{n} \binom{n}{k} (-1)^k \overline{x^k} (-1)^{n-k} \overline{x_T^{n-k}}$$

Thus

$$\overline{x_T^n} = q \overline{x^n} + p \sum_{k=0}^{n} \binom{n}{k} \overline{x^k}\, \overline{x_T^{n-k}}$$

or

$$\overline{x_T^n} = q \overline{x^n} + p \overline{x_T^n} + p \sum_{k=1}^{n} \binom{n}{k} \overline{x^k}\, \overline{x_T^{n-k}}$$

Therefore

$$\overline{x_T^n} = \overline{x^n} + \frac{p}{q} \sum_{k=1}^{n} \binom{n}{k} \overline{x^k}\, \overline{x_T^{n-k}}$$

(d) In determining the number in the system, we may assume (with impunity) that a customer cycles back directly into service instead of to the tail of the queue. This is allowed due to the "memoryless" selection of a new service time each time a customer returns in addition to the independence of the feedback decision. Thus we may consider our queue as an M/G/1 system with $B_T^*(s)$ as the transform for service time, and so Eq. (1.104) may be applied to determine $Q_T(z)$. [Note: This will not work for the determination of $W_T^*(s)$.]

Thus

$$Q_T(z) = B_T^*(\lambda - \lambda z) \frac{(1 - \rho_T)(1 - z)}{B_T^*(\lambda - \lambda z) - z}$$

where

$$B_T^*(s) = \frac{qB^*(s)}{1 - pB^*(s)}$$

and

$$\rho_T = \lambda \overline{x_T^1} = \frac{\lambda \bar{x}}{q}$$

So

$$Q_T(z) = \frac{qB^*(\lambda - \lambda z)}{1 - pB^*(\lambda - \lambda z)} \cdot \frac{\left(1 - \dfrac{\lambda \bar{x}}{q}\right)(1 - z)}{\dfrac{qB^*(\lambda - \lambda z)}{1 - pB^*(\lambda - \lambda z)} - z}$$

$$Q_T(z) = \left(1 - \frac{\lambda \bar{x}}{q}\right) \frac{q(1 - z)B^*(\lambda - \lambda z)}{(q + pz)B^*(\lambda - \lambda z) - z}$$

(e) For M/G/1, $\bar{N} = \bar{q} = \rho + \lambda^2 \overline{x^2}/2(1 - \rho)$. Thus

$$\bar{N} = \rho_T + \frac{\lambda^2 \overline{x_T^2}}{2(1 - \rho_T)}$$

From our earlier results, we have

$$\bar{N} = \frac{\lambda \bar{x}}{q} + \frac{\lambda^2 \left(\dfrac{\overline{x^2}}{q} + 2p\dfrac{(\bar{x})^2}{q^2}\right)}{2\left(1 - \dfrac{\lambda \bar{x}}{q}\right)}$$

$$= \frac{2\lambda \bar{x}(q - \lambda \bar{x}) + \lambda^2[q \overline{x^2} + 2p(\bar{x})^2]}{2q(q - \lambda \bar{x})}$$

$$= \frac{2\lambda \bar{x} q + \lambda^2 q \overline{x^2} - 2\lambda^2(\bar{x})^2(1 - p)}{2q(q - \lambda \bar{x})}$$

Therefore

$$\bar{N} = \frac{2\lambda \bar{x}(1 - \lambda \bar{x}) + \lambda^2 \overline{x^2}}{2(q - \lambda \bar{x})} \qquad \square$$

PROBLEM 5.22

Consider a first-come-first-serve M/G/1 queue with the following changes. The server serves the queue as long as someone is in the system. Whenever the system empties the server goes away on vacation for a certain length of time, which may be a random variable. At the end of his vacation the server returns and begins to serve customers again; if he returns to an empty system then he goes away on vacation again. Let $F(z) = \sum_{j=1}^{\infty} f_j z^j$ be the z-transform for $\tilde{f}$, the number of customers awaiting service when the server returns from vacation to find at least one customer waiting (i.e., f_j is the probability that at the initiation of a busy period the server finds j customers awaiting service).

(a) Derive an expression that gives q_{n+1} in terms of q_n, v_{n+1}, and $\tilde{f}$ (the number of customer arrivals during the server's vacation).

(b) Derive an expression for $Q(z)$, where $Q(z) = \lim E[z^{q_n}]$ as $n \to \infty$ in terms of p_0 (equal to the probability that a departing customer leaves 0 customers behind).

(c) Show that $p_0 = (1 - \rho)/F^{(1)}(1)$, where $F^{(1)}(1) = \partial F(z)/\partial z|_{z=1}$ and $\rho = \lambda \bar{x}$.

(d) Assume now that the service vacation will end whenever a new customer enters the empty system. For this case find $F(z)$ and show that when we substitute it back into our answer for (b) then we arrive at the classical M/G/1 solution.

SOLUTION

(a) Clearly, if $q_n > 0$, then $q_{n+1} = q_n - 1 + v_{n+1}$ as for the usual M/G/1 system. If $q_n = 0$ the server goes on vacation and does not again begin serving until there are $\tilde{f} \geq 1$ in the system [recall that $f_j \Leftrightarrow F(z)$]. Thus $q_{n+1} = \tilde{f} - 1 + v_{n+1}$ for $q_n = 0$. Define a random variable

$$\Delta_{\tilde{f}.k} \stackrel{\Delta}{=} \begin{cases} k & k > 0 \\ \tilde{f} & k = 0 \end{cases}$$

Then

$$q_{n+1} = v_{n+1} - 1 + \Delta_{\tilde{f}.q_n}$$

(b) We have

$$Q_{n+1}(z) = E\left[z^{q_n - 1}\right] = E\left[z^{v_{n+1} - 1 + \Delta_{\tilde{f}.q_n}}\right]$$

$$= E\left[z^{v_{n+1} - 1}\right] E\left[z^{\Delta_{\tilde{f}.q_n}}\right]$$

by independence of v_{n+1}, $\tilde{f}$, and q_n. Letting $n \to \infty$ we find

$$Q(z) = \frac{V(z)}{z} E\left[z^{\Delta_{\tilde{f}.q}}\right]$$

But

$$E\left[z^{\Delta_{\tilde{i}\tilde{q}}}\right] = E\left[z^{\tilde{j}}\right]P[\tilde{q} = 0] + \sum_{k=1}^{\infty} z^k P[\tilde{q} = k]$$

$$= F(z)p_0 + [Q(z) - p_0]$$

Thus

$$Q(z) = \frac{V(z)}{z}\Big[[F(z) - 1]p_0 + Q(z)\Big]$$

or

$$Q(z) = V(z)\frac{p_0[1 - F(z)]}{V(z) - z}$$

[Note that this equation, and its derivation, is the same as that used in the proof of Problem 5.12(b). Here, $V(z) = B^*(\lambda - \lambda z)$, whereas $V(z) = B^*(\lambda - \lambda G(z))$ in 5.12(b). Thus, for $Q(z)$, this vacation problem is equivalent to an M/G/1 bulk arrival system, where bulks may only arrive at the initiation of a busy period.] Finally,

$$Q(z) = B^*(\lambda - \lambda z)\frac{p_0[1 - F(z)]}{B^*(\lambda - \lambda z) - z}$$

(c) From part (b) we have

$$\frac{Q(z)}{V(z)} = \frac{p_0[1 - F(z)]}{V(z) - z}$$

To determine p_0 we evaluate the above equation at $z = 1$ (using L'Hospital's rule). Then

$$1 = \frac{Q(1)}{V(1)} = \lim_{z \to 1} \frac{p_0[1 - F(z)]}{B^*(\lambda - \lambda z) - z}$$

$$= \frac{p_0[-F^{(1)}(1)]}{B^{*(1)}(0)(-\lambda) - 1} = \frac{p_0 F^{(1)}(1)}{1 - \lambda \bar{x}}$$

So

$$p_0 = \frac{1 - \rho}{F^{(1)}(1)}$$

where $\rho = \lambda \bar{x}$.

(d) In this case, $f_1 = 1$ and $f_k = 0$ for $k > 1$. Thus $F(z) = z$. Hence $F^{(1)}(1) = 1$ and $p_0 = 1 - \rho$. So

$$Q(z) = B^*(\lambda - \lambda z)\frac{(1 - \rho)(1 - z)}{B^*(\lambda - \lambda z) - z}$$

which is the P-K transform equation for M/G/1 as given in Eq. (1.104).   □

PROBLEM 5.23

We recognize that an arriving customer who finds k others in the system is delayed by the remaining service time for the customer in service plus the sum of $(k - 1)$ complete service times.

(a) Using the method of supplementary variables as in Problem 5.7, show that we may express the transform of the waiting time pdf as

$$W^*(s) = p_0 + \int_0^\infty \sum_{k=1}^\infty p_k(x_0)[B^*(s)]^{k-1}$$

$$\times \int_0^\infty e^{-sy} r(y + x_0) e^{-\int_0^{y+x_0} r(u)\,du}\,dy$$

$$\times e^{\int_0^{x_0} r(u)\,du}\,dx_0$$

(b) Show that the expression in (a) reduces to $W^*(s)$ as given in Eq. (1.107).

SOLUTION

(a) $W^*(s) = E[e^{-s\tilde{w}}]$, where $\tilde{w}$ is the waiting time of a customer. Since $P[\tilde{w} = 0] = p_0$ (Poisson arrivals) we have $W^*(s) = p_0 + E[e^{-s\tilde{w}} \mid \tilde{w} > 0]$. Assuming a customer has to wait, his waiting time is the sum of the residual service time of the customer being served plus the service times of all those in queue. Thus, if he arrives to find k in the system, his wait is $\tilde{x}_1 + \cdots + \tilde{x}_{k-1} + \tilde{r}$, where the $\tilde{x}_i$ have pdf $b(x)$ with transform $B^*(s)$ and $\tilde{r}$ is the residual life of the customer in service. Given that this customer has received x_0 seconds already, we find

$$P[y < \tilde{r} \le y + \Delta y \mid x_0] = \frac{b(y + x_0)}{1 - B(x_0)}\,\Delta y$$

So

$$E[e^{-s\tilde{w}} \mid k, x_0, \tilde{w} > 0] = E[e^{-s(\tilde{x}_1 + \cdots + \tilde{x}_{k-1} + \tilde{r})} \mid x_0, \tilde{w} > 0]$$

$$= E[e^{-s\tilde{x}_1}] \cdots E[e^{-s\tilde{x}_{k-1}}] \cdot E[e^{-s\tilde{r}} \mid x_0, \tilde{w} > 0]$$

$$= [B^*(s)]^{k-1} \int_0^\infty e^{-sy} \frac{b(y + x_0)}{1 - B(x_0)}\,dy$$

Unconditioning on k and x_0 gives

$$E[e^{-s\tilde{w}} \mid \tilde{w} > 0] = \int_0^\infty \sum_{k=1}^\infty p_k(x_0)[B^*(s)]^{k-1} \int_0^\infty e^{-sy} \frac{b(y + x_0)}{1 - B(x_0)}\,dy\,dx_0$$

Recalling from Problem 5.7(e) that

$$b(y + x_0) = r(y + x_0)e^{-\int_0^{y+x_0} r(u)\,du}$$

$$1 - B(x_0) = e^{-\int_0^{x_0} r(u)\,du}$$

we have the desired equation

$$W^*(s) = p_0 + \int_0^\infty \sum_{k=1}^\infty p_k(x_0)[B^*(s)]^{k-1}$$

$$\times \int_0^\infty e^{-sy} r(y + x_0)e^{-\int_0^{y+x_0} r(u)\,du}\,dy$$

$$\times e^{\int_0^{x_0} r(u)\,du}\,dx_0$$

(b) We recall from part (a) that

$$W^*(s) = p_0 + \int_0^\infty \sum_{k=1}^\infty p_k(x_0)[B^*(s)]^{k-1}$$

$$\times \int_0^\infty e^{-sy} b(y + x_0)\,dy \frac{1}{1 - B(x_0)}\,dx_0$$

To simplify further calculations we define a function

$$F(z) = \int_0^\infty \sum_{k=1}^\infty p_k(x_0)z^k \int_0^\infty e^{-sy} b(y + x_0)\,dy \frac{1}{1 - B(x_0)}\,dx_0$$

Note that

$$W^*(s) = p_0 + \frac{F(B^*(s))}{B^*(s)}$$

and so we must determine $F(z)$. Recalling from Problem 5.7(d) that

$$R(z, x_0) \stackrel{\Delta}{=} \sum_{k=1}^\infty p_k(x_0)z^k$$

we have

$$F(z) = \int_0^\infty R(z, x_0) \int_0^\infty e^{-sy} b(y + x_0)\,dy \frac{1}{1 - B(x_0)}\,dx_0$$

By Problem 5.7(e), we have

$$R(z, x_0) = R(z, 0)e^{-\lambda x_0(1 - z)\,-\,\int_0^{x_0} r(y)\,dy}$$

$$= R(z, 0)e^{-\lambda x_0(1 - z)}[1 - B(x_0)]$$

Thus

$$F(z) = R(z,0) \int_0^\infty e^{-\lambda x_0(1-z)} \int_0^\infty e^{-sy} b(y + x_0) \, dy \, dx_0$$

$$= R(z,0) \int_0^\infty e^{-\lambda x_0(1-z)} \int_{x_0}^\infty e^{-s(t-x_0)} b(t) \, dt \, dx_0$$

Interchanging the order of integration gives

$$F(z) = R(z,0) \int_0^\infty e^{-st} b(t) \int_0^t e^{-x_0(\lambda - \lambda z - s)} \, dx_0 \, dt$$

Carrying out the inner integration yields

$$F(z) = R(z,0) \int_0^\infty e^{-st} b(t) \left[\frac{e^{-t(\lambda - \lambda z - s)} - 1}{s - \lambda(1 - z)} \right] dt$$

and so

$$F(z) = R(z,0) \frac{\left[\int_0^\infty e^{-t(\lambda - \lambda z)} b(t) \, dt - \int_0^\infty e^{-st} b(t) \, dt \right]}{s - \lambda(1 - z)}$$

or

$$F(z) = R(z,0) \frac{B^*(\lambda - \lambda z) - B^*(s)}{s - \lambda(1 - z)}$$

Problem 5.7(e) gives

$$R(z,0) = \frac{\lambda z(z - 1) p_0}{z - B^*(\lambda - \lambda z)}$$

Thus

$$F(z) = p_0 \frac{\lambda z(z - 1)}{s - \lambda(1 - z)} \cdot \frac{B^*(\lambda - \lambda z) - B^*(s)}{z - B^*(\lambda - \lambda z)}$$

Evaluating F at the point $B^*(s)$ gives

$$F(B^*(s)) = p_0 \frac{\lambda B^*(s)[1 - B^*(s)]}{s - \lambda[1 - B^*(s)]}$$

Using

$$W^*(s) = p_0 + \frac{F(B^*(s))}{B^*(s)}$$

we find

$$W^*(s) = p_0 + p_0 \frac{\lambda[1 - B^*(s)]}{s - \lambda[1 - B^*(s)]}$$

$$= p_0 \frac{1}{1 - \frac{\lambda}{s}[1 - B^*(s)]}$$

As $p_0 = 1 - \rho$, we finally obtain

$$W^*(s) = \frac{1 - \rho}{1 - \rho \left[\dfrac{1 - B^*(s)}{s\bar{x}} \right]}$$

which is Eq. (1.107).   □

PROBLEM 5.24

Let us relate $\overline{s^k}$, the kth moment of the time in system to $\overline{N^k}$, the kth moment of the number in system for M/G/1. By analogy to the equation $V(z) = B^*(\lambda - \lambda z)$ derived in Problem 5.11, it follows that $Q(z) = S^*(\lambda - \lambda z)$.

(a) Show that this last expression leads directly to Little's result, namely,

$$\overline{N} = \lambda \bar{s} \overset{\Delta}{=} \lambda T$$

(b) Further, establish the second-moment relationship

$$\overline{N^2} - \overline{N} = \lambda^2 \overline{s^2}$$

(c) Prove that the general relationship is

$$\overline{N(N - 1) \cdots (N - k + 1)} = \lambda^k \overline{s^k}$$

SOLUTION

We will use the stated result

$$Q(z) = S^*(\lambda - \lambda z)$$

(a) We know that $\overline{N} = \bar{q} = Q^{(1)}(1)$. Since

$$Q^{(1)}(z) = S^{*(1)}(\lambda - \lambda z)(-\lambda)$$

then

$$\overline{N} = S^{*(1)}(0)(-\lambda) = (-\bar{s})(-\lambda) = \lambda \bar{s} = \lambda T$$

(b) In a similar fashion we have

$$Q^{(2)}(z) = S^{*(2)}(\lambda - \lambda z)(-\lambda)^2$$

Thus

$$\overline{N^2} - \overline{N} = \overline{q^2} - \bar{q} = Q^{(2)}(1) = S^{*(2)}(0)(-\lambda)^2$$
$$\overline{N^2} - \overline{N} = (-1)^2 \overline{s^2} \lambda^2 = \lambda^2 \overline{s^2}$$

(c) In general

$$Q^{(k)}(1) = S^{*(k)}(0)(-\lambda)^k = (-1)^k \overline{s^k}(-\lambda)^k = \lambda^k \overline{s^k}$$

But $Q(z) = \sum_{j=0}^{\infty} p_j z^j = P(z)$ and so

$$Q^{(k)}(z) = \sum_{j=k}^{\infty} j(j-1)\cdots(j-k+1)p_j z^{j-k}$$

Thus

$$Q^{(k)}(1) = \sum_{j=k}^{\infty} j(j-1)\cdots(j-k+1)p_j$$

$$= \overline{N(N-1)\cdots(N-k+1)}$$

So

$$\overline{N(N-1)\cdots(N-k+1)} = \lambda^k \overline{s^k} \qquad\qquad \square$$

CHAPTER 6

THE QUEUE G/M/m

PROBLEM 6.1

Prove Eq. (1.136). [HINT: Condition on an interarrival time of duration t and then further condition on the time ($\leq t$) it will take to empty the queue.]

SOLUTION

We wish to find p_{ij}, where $j < m < i + 1$, by conditioning on the interarrival time t_{n+1} and also on $\tilde{y} \overset{\Delta}{=}$ time for the queue to empty. We note that the queue becomes empty when $i + 1 - m$ have been served. Thus $\tilde{y} =$ time to serve $i + 1 - m$, which is the sum of $i + 1 - m$ intervals each exponentially distributed with parameter $m\mu$ (recall that the residual life of an exponential random variable is distributed the same as the original interval). Thus $\tilde{y}$ has an $(i + 1 - m)$-stage Erlangian distribution with parameter $m\mu$. Therefore Eq. (1.27) gives the density of $\tilde{y}$ as

$$f_{\tilde{y}}(y) = \frac{m\mu(m\mu y)^{i-m}}{(i-m)!} e^{-m\mu y}$$

For the remaining interval, we note that $m - j$ of the m customers that are left must be served in $t_{n+1} - \tilde{y}$ seconds. Given that $t_{n+1} = t$ and $\tilde{y} = y$, this last occurs with probability

$$\binom{m}{j} \left[1 - e^{-\mu(t-y)}\right]^{m-j} \left[e^{-\mu(t-y)}\right]^{j}$$

Removing the condition on $\bar{y}$ gives

$$p_{ij}|_{t_{n+1}=t} = \int_0^t \binom{m}{j} e^{-\mu(t-y)j} \left[1 - e^{-\mu(t-y)}\right]^{m-j} f_{\bar{y}}(y)\,dy$$

$$p_{ij}|_{t_{n+1}=t} = \binom{m}{j} e^{-\mu t j} \int_0^t e^{\mu y j} \left[1 - e^{-\mu(t-y)}\right]^{m-j} \frac{m\mu(m\mu y)^{i-m}}{(i-m)!} e^{-m\mu y}\,dy$$

$$p_{ij}|_{t_{n+1}=t} = \binom{m}{j} e^{-\mu t j} \int_0^t \frac{m\mu(m\mu y)^{i-m}}{(i-m)!} e^{-\mu y(m-j)} \left[1 - e^{-\mu(t-y)}\right]^{m-j}\,dy$$

Finally, removing the condition on t_{n+1} gives

$$p_{ij} = \int_0^\infty \binom{m}{j} e^{-j\mu t} \left[\int_0^t \frac{(m\mu y)^{i-m}}{(i-m)!} (e^{-\mu y} - e^{-\mu t})^{m-j} m\mu\,dy\right] dA(t) \qquad \square$$

PROBLEM 6.2

Consider $E_2/M/1$ (with infinite queueing room).

(a) Solve for r_k in terms of σ.
(b) Evaluate σ explicitly.

SOLUTION

(a) By Eq. (1.125) for G/M/1

$$r_k = (1 - \sigma)\sigma^k \qquad k = 0, 1, 2, \dots$$

(b) Equation (1.124) gives $\sigma = A^*(\mu - \mu\sigma)$, where $0 < \sigma < 1$. Now $A^*(s) = [2\lambda/(s + 2\lambda)]^2$ for $E_2/M/1$. So $\sigma = [2\lambda/(\mu - \mu\sigma + 2\lambda)]^2$ or

$$\sigma[\mu^2(1 - 2\sigma + \sigma^2) + 4\lambda\mu(1 - \sigma) + 4\lambda^2] = 4\lambda^2$$

$$\mu^2\sigma^3 - (2\mu^2 + 4\lambda\mu)\sigma^2 + (\mu^2 + 4\lambda\mu + 4\lambda^2)\sigma - 4\lambda^2 = 0$$

Since $\sigma = 1$ is always a root, we have

$$(\sigma - 1)[\mu^2\sigma^2 - \mu(\mu + 4\lambda)\sigma + 4\lambda^2] = 0$$

We now seek that root of the quadratic

$$\sigma^2 - \left(1 + \frac{4\lambda}{\mu}\right)\sigma + 4\left(\frac{\lambda}{\mu}\right)^2 = 0$$

for which $0 < \sigma < 1$. Since $\rho = \lambda/\mu$, then $\sigma^2 - (1 + 4\rho)\sigma + 4\rho^2 = 0$. Thus

$$\sigma = \frac{1 + 4\rho \pm \sqrt{1 + 8\rho}}{2}$$

Since $\rho \geq 0$, then $(1 + 4\rho + \sqrt{1 + 8\rho})/2 \geq 1$. Therefore the root we seek is

$$\sigma = \frac{1 + 4\rho - \sqrt{1 + 8\rho}}{2}$$

Let us show that $0 < \sigma < 1$. For $\rho > 0$, we clearly have

$$\sigma > \frac{1 + 4\rho - \sqrt{1 + 8\rho + 16\rho^2}}{2}$$

or $\sigma > 0$. Also, for $\rho < 1$ we have $16\rho^2 < 16\rho$. Hence

$$16\rho^2 - 8\rho + 1 < 16\rho - 8\rho + 1 = 8\rho + 1$$

and so

$$4\rho - 1 \leq |4\rho - 1| < \sqrt{1 + 8\rho}$$

Thus $1 + 4\rho - \sqrt{1 + 8\rho} < 2$, which yields $\sigma < 1$. For a stable system we have $0 < \rho < 1$, which, from the above argument, implies $0 < \sigma < 1$.   □

PROBLEM 6.3

Consider M/M/m.

(a) How do p_k and r_k compare?
(b) For G/M/m, find $P[\text{arrival queues}]$ in terms of J and σ. Compare this result to the Erlang-C formula (see Section 1.5) for $P[\text{arrival queues}]$ for the M/M/m system.

SOLUTION

(a) For Poisson arrivals, we know $p_k = r_k$ (see Section 1.3).
(b) Equation (1.130) gives

$$P[\text{arrival queues}] = \sum_{k=m}^{\infty} r_k = \frac{J\sigma}{1 - \sigma} \quad \text{(G/M/m)}$$

The Erlang-C formula gives

$$P[\text{queueing}] = \sum_{k=m}^{\infty} p_k = p_0 \left(\frac{(m\rho)^m}{m!} \right) \left(\frac{1}{1 - \rho} \right)$$

Since $p_k = r_k$ for all k, these two probabilities are the same.
    [If we wish to solve for the constant J, we proceed as follows:

$$\sigma = A^*(m\mu - m\mu\sigma) = \frac{\lambda}{m\mu(1 - \sigma) + \lambda}$$

and so $\sigma = \lambda/m\mu = \rho$. Equating the above probabilities we obtain $J = p_0(m\rho)^m/\rho m!$ where p_0 is given in Eq. (1.87).]   □

PROBLEM 6.4

Prove Eq. (1.127).

SOLUTION

We have $W = \int_0^\infty y\,dW(y)$. Using Eq. (1.126) to determine $dW(y)$,

$$W = \int_0^x y\sigma\mu(1 - \sigma)e^{-\mu(1-\sigma)y}\,dy = \sigma\,\frac{1}{\mu(1 - \sigma)}$$

which is Eq. (1.127).   □

PROBLEM 6.5

Consider an $H_2/M/1$ system in which $\lambda_1 = 2$, $\lambda_2 = 1$, $\mu = 2$, and $\alpha_1 = 5/8$ (here λ_i takes the role of μ_i for this H_2 arrival distribution).

(a) Find σ.
(b) Find r_k.
(c) Find $w(y)$.
(d) Find W.

SOLUTION

(a) For $H_2/M/1$, $A^*(s) = \alpha_1\lambda_1/(s + \lambda_1) + \alpha_2\lambda_2/(s + \lambda_2)$, where $\alpha_2 = 1 - \alpha_1$. Thus

$$A^*(s) = \frac{5}{8} \cdot \frac{2}{s + 2} + \frac{3}{8} \cdot \frac{1}{s + 1}$$

Now $\sigma = A^*(\mu - \mu\sigma)$ for G/M/1, so using $\mu = 2$ we find

$$\sigma = \frac{5}{8} \cdot \frac{2}{2 - 2\sigma + 2} + \frac{3}{8} \cdot \frac{1}{2 - 2\sigma + 1}$$

or

$$\sigma(4 - 2\sigma)(3 - 2\sigma) = \frac{21}{4} - \frac{13}{4}\sigma$$

Thus

$$4\sigma^3 - 14\sigma^2 + \frac{61}{4}\sigma - \frac{21}{4} = 0$$

Since $\sigma = 1$ is always a root, we factor

$$(\sigma - 1)\left(4\sigma^2 - 10\sigma + \frac{21}{4}\right) = 0$$

or

$$(\sigma - 1)(4\sigma - 7)\left(\sigma - \frac{3}{4}\right) = 0$$

The condition $0 < \sigma < 1$ gives

$$\sigma = \tfrac{3}{4}$$

(b) For G/M/1, $r_k = (1 - \sigma)\sigma^k$. So

$$r_k = \tfrac{1}{4}\left(\tfrac{3}{4}\right)^k$$

(c) Differentiating Eq. (1.126) gives

$$w(y) = (1 - \sigma)u_0(y) + \sigma\mu(1 - \sigma)e^{-\mu(1-\sigma)y} \qquad y \geq 0$$

Thus

$$w(y) = \frac{1}{4}u_0(y) + \frac{3}{8}e^{-\frac{y}{2}} \qquad y \geq 0$$

(d) By Eq. (1.127) $W = \sigma/\mu(1 - \sigma)$. Thus

$$W = \frac{3/4}{2(1/4)} = \frac{3}{2} \qquad\qquad \square$$

PROBLEM 6.6

Consider a D/M/1 system with $\mu = 2$ and with the same ρ as in the previous problem.

(a) Find σ (correct to two decimal places).
(b) Find r_k.
(c) Find $w(y)$.
(d) Find W.

SOLUTION

(a) Since ρ and $\bar{x}\,(= 1/\mu)$ are the same as in Problem 6.5, so is $\bar{t}$. Thus

$$\bar{t} = \frac{\alpha_1}{\lambda_1} + \frac{\alpha_2}{\lambda_2} = \frac{5}{16} + \frac{6}{16} = \frac{11}{16}$$

For D/M/1, $A^*(s) = e^{-s\bar{t}}$ and so $A^*(s) = e^{-\frac{11}{16}s}$. Now

$$\sigma = A^*(\mu - \mu\sigma) = A^*(2 - 2\sigma) = e^{-\frac{11}{8}(1-\sigma)}$$

or

$$\log_e \sigma = \frac{11}{8}(\sigma - 1)$$

Since $\sigma - 1$ is tangent to $\log_e \sigma$ at $\sigma = 1$, we see that this last equation has exactly two roots.

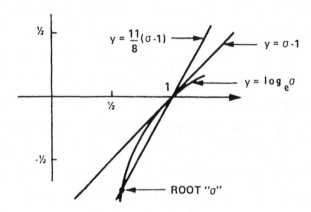

Disregarding the root at $\sigma = 1$, we find a root satisfying $0 < \sigma < 1$ numerically to obtain

$$\sigma \cong 0.51$$

(b) For G/M/1, $r_k = (1 - \sigma)\sigma^k$. So

$$r_k \cong (0.49)(0.51)^k$$

(c) From Eq. (1.126) we have

$$w(y) = (1 - \sigma)u_0(y) + \sigma\mu(1 - \sigma)e^{-\mu(1-\sigma)y} \qquad y \geq 0$$

Since $\mu = 2$ and $\sigma \cong 0.51$ we have

$$w(y) \cong (0.49)u_0(y) + (0.50)e^{-(0.98)y} \qquad y \geq 0$$

(d) From Eq. (1.127) we have $W = \sigma/\mu(1 - \sigma)$. Thus

$$W \cong 0.52 \qquad\qquad \square$$

PROBLEM 6.7

Consider a G/M/1 queueing system with room for at most two customers (one in service plus one waiting). Find r_k ($k = 0, 1, 2$) in terms of μ and $A^*(s)$.

SOLUTION

Our task is to solve $\mathbf{r} = \mathbf{rP}$, where $\mathbf{r} = [r_0, r_1, r_2]$. First we must find the 3×3 matrix

$$\mathbf{P} = \begin{bmatrix} p_{00} & p_{01} & p_{02} \\ p_{10} & p_{11} & p_{12} \\ p_{20} & p_{21} & p_{22} \end{bmatrix}$$

Clearly $p_{02} = 0$. Furthermore, we now show that the last two rows of $\mathbf{P}$ are equal. If C_n arrives to find a full system ($q'_n = 2$) then he will be lost, whereas if he arrives to find one in the system ($q'_n = 1$) then he queues up and makes a full system. Due to the memoryless property of the service time distribution, these two cases are equivalent in terms of what the next customer, C_{n+1}, sees. Thus

$$p_{2i} = p_{1i} \qquad \text{for } i = 0, 1, 2$$

and so

$$\mathbf{P} = \begin{bmatrix} p_{00} & p_{01} & 0 \\ p_{10} & p_{11} & p_{12} \\ p_{10} & p_{11} & p_{12} \end{bmatrix}$$

Let us now find these matrix entries:

$$p_{01} = p_{12} = P[\text{service time} > \text{interarrival time}] = \int_0^\infty e^{-\mu t} \, dA(t) = A^*(\mu)$$

Thus also $p_{00} = 1 - p_{01} = 1 - A^*(\mu)$. To find p_{10} assume that $q'_n = 1$ (thus there will be 2 in the system) and that $q'_{n+1} = 0$. Thus we have

$$p_{10} = P[\text{residual life of } C_{n-1} \text{ plus service time of } C_n \le \text{interarrival time}]$$

Since the service time is exponential, this sum is Erlangian; it has density $\mu(\mu x)e^{-\mu x}$ and distribution $1 - e^{-\mu x} - \mu x e^{-\mu x}$. Thus

$$p_{10} = \int_0^\infty (1 - e^{-\mu t} - \mu t e^{-\mu t}) \, dA(t)$$

or

$$p_{10} = 1 - A^*(\mu) + \mu A^{*(1)}(\mu)$$

Finally

$$p_{11} = 1 - p_{10} - p_{12} = -\mu A^{*(1)}(\mu)$$

Thus

$$\mathbf{P} = \begin{bmatrix} 1 - A^*(\mu) & A^*(\mu) & 0 \\ 1 - A^*(\mu) + \mu A^{*(1)}(\mu) & -\mu A^{*(1)}(\mu) & A^*(\mu) \\ 1 - A^*(\mu) + \mu A^{*(1)}(\mu) & -\mu A^{*(1)}(\mu) & A^*(\mu) \end{bmatrix}$$

Now we determine r_k. From $\mathbf{r} = \mathbf{r}\mathbf{P}$ ($[r_0, r_1, r_2] = [r_0, r_1, r_2]\mathbf{P}$) we have the two independent balance equations

$$r_1 = r_0 A^*(\mu) - (r_1 + r_2)\mu A^{*(1)}(\mu)$$

$$r_2 = (r_1 + r_2)A^*(\mu)$$

or, since $r_1 + r_2 = 1 - r_0$,

$$r_1 = r_0 A^*(\mu) - (1 - r_0)\mu A^{*(1)}(\mu)$$

$$r_2 = (1 - r_0)A^*(\mu)$$

The conservation of probability equation $r_0 + r_1 + r_2 = 1$ now yields

$$r_0 + \left[r_0 A^*(\mu) - (1 - r_0)\mu A^{*(1)}(\mu)\right] + \left[(1 - r_0)A^*(\mu)\right] = 1$$

Thus

$$r_0 = \frac{1 - A^*(\mu) + \mu A^{*(1)}(\mu)}{1 + \mu A^{*(1)}(\mu)}$$

Using this result we also find

$$r_1 = \frac{\left[1 - A^*(\mu)\right]A^*(\mu)}{1 + \mu A^{*(1)}(\mu)}$$

$$r_2 = \frac{\left[A^*(\mu)\right]^2}{1 + \mu A^{*(1)}(\mu)} \qquad\qquad \square$$

PROBLEM 6.8

Consider a G/M/1 system in which the cost of making a customer wait y sec is

$$c(y) = ae^{by}$$

(a) Find the average cost of queueing for a customer.
(b) Under what conditions will the average cost be finite?

SOLUTION

For G/M/1,

$$W(y) = 1 - \sigma e^{-\mu(1-\sigma)y} \qquad y \geq 0$$

Thus

$$w(y) = (1 - \sigma)u_0(y) + \sigma\mu(1 - \sigma)e^{-\mu(1-\sigma)y} \qquad y \geq 0$$

(a) The average cost of queueing is

$$\bar{c} \stackrel{\Delta}{=} \int_0^\infty c(y)\, dW(y) = \int_0^\infty a e^{by} w(y)\, dy$$

$$= a(1 - \sigma) + \int_0^\infty a\sigma\mu(1 - \sigma)e^{-[\mu(1-\sigma)-b]y}\, dy$$

Assuming $\mu(1 - \sigma) - b > 0$ so that the integral exists, we find

$$\bar{c} = a(1 - \sigma) + \frac{a\sigma\mu(1 - \sigma)}{\mu(1 - \sigma) - b}$$

or

$$\bar{c} = a(1 - \sigma)\frac{\mu - b}{\mu(1 - \sigma) - b}$$

(b) The average cost is finite if

$$b < \mu(1 - \sigma)$$

as indicated in part (a). □

CHAPTER 7

THE QUEUE G/G/1

PROBLEM 7.1

From Eq. (1.142) show that $C^*(s) = A^*(-s)B^*(s)$.

SOLUTION

From

$$c(u) = \int_0^\infty b(u + t)a(t)\, dt$$

we form the Laplace transform as

$$C^*(s) \triangleq \int_{-\infty}^\infty c(u)e^{-su}\, du$$

$$= \int_{-\infty}^\infty \left(\int_0^\infty b(u + t)a(t)\, dt \right) e^{-su}\, du$$

$$= \int_0^\infty a(t)e^{st} \left(\int_{-\infty}^\infty b(u + t)e^{-s(u+t)}\, du \right) dt$$

Let $x = u + t$, $dx = du$, which yields

$$C^*(s) = \int_0^\infty a(t)e^{st} \left(\int_{-\infty}^\infty b(x)e^{-sx}\, dx \right) dt$$

Since $b(x) = 0$ for $x < 0$,

$$C^*(s) = \int_0^\infty a(t)e^{st} \left(\int_0^\infty b(x)e^{-sx}\,dx \right) dt$$

$$C^*(s) = \int_0^\infty a(t)e^{st}\,dt \int_0^\infty b(x)e^{-sx}\,dx$$

$$C^*(s) = A^*(-s)B^*(s)$$

□

PROBLEM 7.2

Find $C(u)$ for M/M/1.

SOLUTION

Integrating Eq. (1.142). we have

$$C(u) = \int_0^\infty B(u + t)a(t)\,dt$$

For M/M/1,

$$B(x) = \begin{cases} 1 - e^{-\mu x} & x \geq 0 \\ 0 & x < 0 \end{cases}$$

and

$$a(t) = \begin{cases} \lambda e^{-\lambda t} & t \geq 0 \\ 0 & t < 0 \end{cases}$$

To determine the limits of integration for $C(u)$, we consider two cases:

Case (1): $u \geq 0$

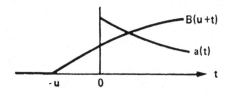

$$C(u) = \int_0^\infty (1 - e^{-\mu(u+t)})\lambda e^{-\lambda t}\,dt$$

$$= \int_0^\infty \lambda e^{-\lambda t}\,dt - \int_0^\infty e^{-\mu u}\lambda e^{-(\lambda + \mu)t}\,dt$$

$$C(u) = 1 - \frac{\lambda}{\lambda + \mu}e^{-\mu u}$$

Case (2): $u < 0$

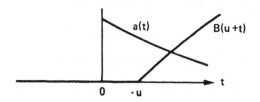

$$C(u) = \int_{-u}^{\infty} (1 - e^{-\mu(u+t)}) \lambda e^{-\lambda t}\, dt$$

$$= \int_{-u}^{\infty} \lambda e^{-\lambda t}\, dt - \int_{-u}^{\infty} e^{-\mu u} \lambda e^{-(\lambda+\mu)t}\, dt$$

$$= e^{\lambda u} - e^{-\mu u}\left(\frac{\lambda}{\lambda + \mu} e^{(\lambda+\mu)u} \right)$$

$$C(u) = \frac{\mu}{\lambda + \mu} e^{\lambda u}$$

So

$$C(u) = \begin{cases} 1 - \dfrac{\lambda}{\lambda + \mu} e^{-\mu u} & u \geq 0 \\[3mm] \dfrac{\mu}{\lambda + \mu} e^{\lambda u} & u < 0 \end{cases}$$

We sketch $C(u)$ as follows:

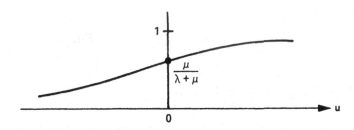

□

PROBLEM 7.3

Consider the system M/D/1 with a fixed service time of $\bar{x}$ sec.

(a) Find

$$C(u) = P[u_n \leq u]$$

and sketch its shape.

(b) Find $E[u_n]$.

SOLUTION

(a) From the definition of u_n [Eq. (1.140)], we have

$$C(u) = P[x_n - t_{n+1} \leq u]$$

Since we have an M/D/1 system, then

$$C(u) = P[\bar{x} - u \leq t_{n+1}]$$
$$= 1 - P[t_{n+1} \leq \bar{x} - u]$$

and so $C(u) = 1 - A(\bar{x} - u)$. We also know that

$$A(t) = \begin{cases} 0 & t < 0 \\ 1 - e^{-\lambda t} & t \geq 0 \end{cases}$$

Thus

$$C(u) = \begin{cases} 1 & \bar{x} < u \\ e^{-\lambda(\bar{x}-u)} & \bar{x} \geq u \end{cases}$$

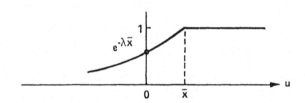

(b) We have

$$E[u_n] = E[x_n - t_{n+1}] = E[x_n] - E[t_{n+1}]$$
$$E[u_n] = \bar{x} - \bar{t} = \bar{t}(\rho - 1)$$

(Note: This is also true for G/G/1.)   □

PROBLEM 7.4

Consider the case where $\rho = 1 - \varepsilon$ for $0 < \varepsilon \ll 1$. Let us expand $W(y - u)$ in Eq. (1.151) as

$$W(y - u) = W(y) - uW^{(1)}(y) + \frac{u^2}{2}W^{(2)}(y) + R(u, y)$$

where $W^{(n)}(y)$ is the nth derivative of $W(y)$ and $R(u, y)$ is such that $\int_{-\infty}^{x} R(u, y)\, dC(u)$ is negligible due to the slow variation of $W(y)$ when $\rho = 1 - \varepsilon$. Let $\overline{u^k}$ denote the kth moment of $\tilde{u}$.

(a) Under these conditions convert Lindley's integral equation to a second-order linear differential equation involving $\overline{u^2}$ and $\bar{u}$.

(b) With the boundary condition $W(0) = 0$, solve the equation found in (a) and express the mean wait W in terms of the first two moments of $\tilde{t}$ and $\tilde{x}$.

SOLUTION

(a) Lindley's integral equation [Eq. (1.151)] is

$$W(y) = \begin{cases} \int_{-\infty}^{y} W(y - u)\, dC(u) & y \geq 0 \\ 0 & y < 0 \end{cases}$$

Integrating both sides of the expansion for $W(y - u)$ given in the problem statement over all u, assuming $\int_{-\infty}^{\infty} R(u, y)\, dC(u)$ is negligible, and noting that

$$W(y) = \int_{-\infty}^{\infty} W(y - u)\, dC(u) \qquad \text{for } y \geq 0$$

[since $W(y - u)$ is zero for $u > y$], we have

$$W(y) = \int_{-\infty}^{\infty} \left[W(y) - uW^{(1)}(y) + \frac{u^2}{2} W^{(2)}(y) \right] dC(u)$$

or

$$W(y) = W(y) - W^{(1)}(y)\bar{u} + W^{(2)}(y)\frac{\overline{u^2}}{2}$$

The required differential equation is therefore

$$\frac{\overline{u^2}}{2} W^{(2)}(y) - \bar{u} W^{(1)}(y) = 0$$

(b) Integrating the above equation yields the first-order equation

$$\frac{\overline{u^2}}{2} W^{(1)}(y) - \bar{u} W(y) = C$$

The homogeneous solution is

$$W_h(y) = K e^{(2\bar{u}/\overline{u^2})y}$$

Substituting back into our differential equation, we see that the particular solution must be

$$W_p(y) = -\frac{C}{\bar{u}}$$

Thus

$$W(y) = K e^{(2\bar{u}/\overline{u^2})y} - \frac{C}{\bar{u}}$$

The boundary condition $W(0) = 0$ gives $C/\bar{u} = K$. Thus

$$W(y) = K(e^{(2\bar{u}/\overline{u^2})y} - 1)$$

Now, as $y \to \infty$ we must have $W(y) \to 1$ (it is a PDF). Since $2\bar{u}/\overline{u^2} < 0$ for a stable system,

$$\lim_{y \to \infty} e^{(2\bar{u}/\overline{u^2})y} = 0$$

and so $K = -1$. Thus

$$W(y) = 1 - e^{(2\bar{u}/\overline{u^2})y} \qquad y \geq 0$$

The mean wait W is simply

$$W = -\frac{\overline{u^2}}{2\bar{u}}$$

Since $\tilde{u} = \tilde{x} - \tilde{t}$, we have that $\bar{u} = \bar{x} - \bar{t} = \bar{t}(\rho - 1)$. Also

$$\overline{u^2} = \overline{(\tilde{x} - \tilde{t})^2} = \overline{x^2} + \overline{t^2} - 2\bar{x}\bar{t}$$

or

$$\overline{u^2} = \sigma_a^2 + \sigma_b^2 + (\bar{x})^2 + (\bar{t})^2 - 2\bar{x}\bar{t}$$

Thus

$$\overline{u^2} = \sigma_a^2 + \sigma_b^2 + (\bar{t})^2(1 - \rho)^2$$

Substituting these expressions into our expression for W, we obtain

$$W = \frac{\sigma_a^2 + \sigma_b^2}{2\bar{t}(1 - \rho)} + \frac{1}{2}\bar{t}(1 - \rho) \qquad (\rho \to 1) \qquad \square$$

PROBLEM 7.5

Consider the D/E$_r$/1 queueing system, with a constant interarrival time (of $\bar{t}$ sec) and a service time pdf given as in Eq. (1.27).

(a) Find $C(u)$.

(b) Show that Lindley's integral equation yields $W(y - \bar{t}) = 0$ for $y < \bar{t}$ and

$$W(y - \bar{t}) = \int_0^y W(y - w)\, dB(w) \qquad \text{for } y \geq \bar{t}$$

(c) Assume the following solution for $W(y)$:

$$W(y) = 1 + \sum_{i=1}^{r} a_i e^{\alpha_i y} \qquad y \geq 0$$

where a_i and α_i may both be complex, but where $\text{Re}(\alpha_i) < 0$ for $i = 1, 2, \ldots, r$. Using this assumed solution, show that the following equations must hold:

$$e^{-\alpha_i \bar{t}} = \left(\frac{r\mu}{r\mu + \alpha_i} \right)^r \qquad i = 1, 2, \ldots, r$$

$$\sum_{i=0}^{r} \frac{a_i}{(r\mu + \alpha_i)^{j+1}} = 0 \qquad j = 0, 1, \ldots, r - 1$$

where $a_0 = 1$ and $\alpha_0 = 0$. Note that $\{\alpha_i\}$ may be found from the first set of (transcendental) equations, and then the second set gives $\{a_i\}$. It can be shown that the α_i are distinct. See [SYSK 60].

SOLUTION

(a) From Problem 7.2 we recall that

$$C(u) = \int_0^{\infty} B(u + t)\, dA(t)$$

For $D/E_r/1$, $dA(t) = a(t)\,dt = u_0(t - \bar{t})\,dt$. So $C(u) = B(u + \bar{t})$. Now Eq. (1.27) gives the service time density as

$$b(x) = \frac{r\mu (r\mu x)^{r-1}}{(r-1)!} e^{-r\mu x} \qquad x \geq 0$$

The corresponding distribution function (obtained through repeated integration by parts) is

$$B(x) = 1 - \sum_{i=0}^{r-1} \frac{(r\mu x)^i}{i!} e^{-r\mu x} \qquad x \geq 0$$

Therefore

$$C(u) = B(u + \bar{t}) = \begin{cases} 0 & u < -\bar{t} \\ 1 - \displaystyle\sum_{i=0}^{r-1} \frac{[r\mu(u + \bar{t})]^i}{i!} e^{-r\mu(u+\bar{t})} & u \geq -\bar{t} \end{cases}$$

(b) Using Eq. (1.151), evaluating it at $y - \bar{t}$, and then recalling that $C(u) = B(u + \bar{t})$, we find

$$W(y - \bar{t}) = \begin{cases} \int_{-\infty}^{y-\bar{t}} W(y - \bar{t} - u)\, dB(u + \bar{t}) & y \geq \bar{t} \\ 0 & y < \bar{t} \end{cases}$$

Substituting $w = u + \bar{t}$ gives

$$W(y - \bar{t}) = \begin{cases} \int_{-x}^{y} W(y - w)\,dB(w) & y \geq \bar{t} \\ 0 & y < \bar{t} \end{cases}$$

as desired.

(c) We first rewrite the assumed solution for $W(y)$ in the form

$$W(y) = \sum_{i=0}^{r} a_i e^{\alpha_i y} \qquad y \geq 0$$

where we define $a_0 = 1$ and $\alpha_0 = 0$. For $y \geq \bar{t}$, we will substitute this assumed solution into the equation from part (b) to yield the desired relations. The left-hand side (LHS) of the equation becomes

$$\text{LHS} = W(y - \bar{t}) = \sum_{i=0}^{r} a_i e^{\alpha_i (y - \bar{t})}$$

The right-hand side (RHS) is

$$\text{RHS} = \int_{0}^{y} W(y - w)\,dB(w) = \int_{0}^{y} \sum_{i=0}^{r} a_i e^{\alpha_i (y - w)}\,dB(w)$$

$$= \sum_{i=0}^{r} a_i e^{\alpha_i y} \int_{0}^{y} e^{-\alpha_i w}\,dB(w)$$

$$\text{RHS} = \sum_{i=0}^{r} a_i e^{\alpha_i y} \int_{0}^{y} \frac{r\mu(r\mu w)^{r-1}}{(r-1)!} e^{-(r\mu + \alpha_i)w}\,dw$$

Repeated integration by parts gives

$$\text{RHS} = \sum_{i=0}^{r} a_i e^{\alpha_i y} \left[-\sum_{j=1}^{r} \frac{(r\mu w)^{r-j}}{(r-j)!} \left(\frac{r\mu}{r\mu + \alpha_i} \right)^{j} e^{-(r\mu + \alpha_i)w} \right] \Bigg|_{0}^{y}$$

$$= \sum_{i=0}^{r} a_i e^{\alpha_i y} \left[\left(\frac{r\mu}{r\mu + \alpha_i} \right)^{r} - \sum_{j=1}^{r} \frac{(r\mu y)^{r-j}}{(r-j)!} \left(\frac{r\mu}{r\mu + \alpha_i} \right)^{j} e^{-(r\mu + \alpha_i)y} \right]$$

$$= \sum_{i=0}^{r} a_i e^{\alpha_i y} \left(\frac{r\mu}{r\mu + \alpha_i} \right)^{r} - \sum_{j=1}^{r} \frac{(r\mu)^{r} y^{r-j}}{(r-j)!} \sum_{i=0}^{r} \frac{a_i e^{-r\mu y}}{(r\mu + \alpha_i)^{j}}$$

Equating coefficients of $e^{\alpha_i y}$ on LHS and RHS gives

$$e^{-\alpha_i \bar{t}} = \left(\frac{r\mu}{r\mu + \alpha_i} \right)^{r} \qquad 1 \leq i \leq r$$

while the coefficient of $e^{-r\mu y}$ must be zero, and so

$$\sum_{j=1}^{r} \frac{r\mu^r y^{r-j}}{(r-j)!} \sum_{i=0}^{r} \frac{a_i}{(r\mu + \alpha_i)^j} = 0 \qquad \text{(for all } y \ge 0)$$

Thus

$$\sum_{i=0}^{r} \frac{a_i}{(r\mu + \alpha_i)^j} = 0 \qquad 1 \le j \le r$$

or

$$\sum_{i=0}^{r} \frac{a_i}{(r\mu + \alpha_i)^{j+1}} = 0 \qquad 0 \le j \le r - 1 \qquad \Box$$

PROBLEM 7.6

Consider the following queueing systems in which *no queue* is permitted. Customers who arrive to find the system busy must leave without service.

(a) M/M/1/1: Solve for $p_k = P[k \text{ in system}]$.
(b) M/H$_2$/1/1: This is the case of Eq. (1.28) with $R = 2$, $\alpha_1 = \alpha$, $\alpha_2 = 1 - \alpha$, $\mu_1 = 2\mu\alpha$, and $\mu_2 = 2\mu(1 - \alpha)$.

   (i) Find the mean service time $\bar{x}$.
   (ii) Solve for p_0 (an empty system), p_α (a customer is in service and is being served at the rate $2\mu\alpha$), and $p_{1-\alpha}$ [a customer is in service and is being served at the rate $2\mu(1 - \alpha)$].

(c) H$_2$/M/1/1: Here $A(t)$ is hyperexponential as in (b), but with parameters $\mu_1 = 2\lambda\alpha$ and $\mu_2 = 2\lambda(1 - \alpha)$ instead. Draw the state-transition-rate diagram (with labels on branches) for the following four states: (i, j) is the state with the "arriving" customer in arrival stage i (arriving at rate μ_i) and j customers in service, $i = 1, 2$ and $j = 0, 1$.
(d) M/E$_r$/1/1: Solve for $P_j = P[j \text{ stages of service left to go}]$.
(e) M/D/1/1: With all service times equal to $\bar{x}$.

   (i) Find the probability of an empty system.
   (ii) Find the fraction of lost customers.

(f) E$_2$/M/1/1: Define the four states as (i, j), where i is the number of "arrival" stages left to go and j is the number of customers in service, $i = 1, 2$ and $j = 0, 1$. Draw the labeled state-transition-rate diagram.

SOLUTION

(a) M/M/1/1: The state-transition-rate diagram is

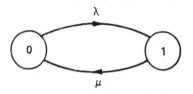

$$\lambda p_0 = \mu p_1 \quad \text{and} \quad p_0 + p_1 = 1$$

Thus

$$p_1 = \frac{\lambda}{\lambda + \mu}$$

$$p_0 = \frac{\mu}{\lambda + \mu}$$

(b) M/H$_2$/1/1: Letting 1_i be the state with one customer in service and that customer in stage i, the state-transtion-rate diagram is

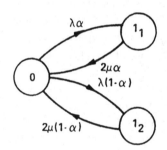

(i) We have

$$\bar{x} = \alpha \left(\frac{1}{2\mu\alpha} \right) + (1 - \alpha) \left(\frac{1}{2\mu(1 - \alpha)} \right) = \frac{1}{\mu}$$

(ii) The balance equations are

$$\lambda \alpha p_0 = 2\mu\alpha p_\alpha$$

$$\lambda(1 - \alpha)p_0 = 2\mu(1 - \alpha)p_{1-\alpha}$$

$$p_0 + p_\alpha + p_{1-\alpha} = 1$$

Thus

$$p_0 = \frac{\mu}{\lambda + \mu}$$

$$p_\alpha = p_{1-\alpha} = \frac{\lambda}{2(\lambda + \mu)}$$

(c) $H_2/M/1/1$: The state-transition-rate diagram is

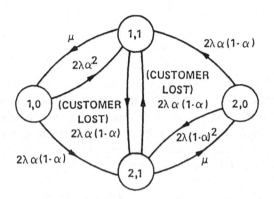

(d) $M/E_r/1/1$: Letting j denote the number of stages of service left, the state-transition-rate diagram is

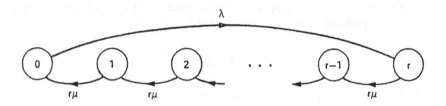

The equilibrium equations for P_j are

$$\lambda P_0 = r\mu P_1$$

$$r\mu P_1 = r\mu P_2$$

$$\vdots$$

$$r\mu P_{r-1} = r\mu P_r$$

These equations plus the conservation of probability equation

$$P_0 + P_1 + \cdots + P_r = 1$$

yield

$$P_0 = \frac{\mu}{\lambda + \mu}$$

$$P_j = \frac{\lambda}{r(\lambda + \mu)} \qquad 1 \le j \le r$$

(e) M/D/1/1

   (i) Note that M/D/1/1 = $\lim_{r \to \infty}$ M/E$_r$/1/1. Then, by part (d),

$$P[\text{empty system}] = \frac{\mu}{\lambda + \mu} = \frac{1/\bar{x}}{\lambda + 1/\bar{x}} = \frac{1}{1 + \lambda\bar{x}}$$

[Note: The probability of an empty system for an M/G/1/1 system may be found by the following renewal theoretic argument. Since we have Poisson arrivals, the time between a departure and a new arrival is exponentially distributed with mean $1/\lambda$. Thus the mean length of an idle period, say, $\bar{I}$, is $1/\lambda$. Since no queue is allowed, the mean length of a busy period, say, $\bar{B}$, is $\bar{x}$. From renewal theory,

$$P[\text{empty system}] = \frac{\bar{I}}{\bar{I} + \bar{B}} = \frac{1/\lambda}{1/\lambda + \bar{x}} = \frac{1}{1 + \lambda\bar{x}}$$

for M/G/1/1. Thus portions of parts (a), (b), (d), and (e) are special cases of this result.]

  (ii) The fraction of lost customers = probability of an arrival to a busy system = probability of a busy system (Poisson arrivals)

$$= 1 - p_0 = \frac{\lambda\bar{x}}{1 + \lambda\bar{x}}$$

(f) E$_2$/M/1/1: The state-transition-rate diagram is

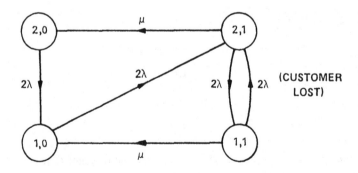

PROBLEM 7.7

Consider a single-server queueing system in which the interarrival time is chosen with probability α from an exponential distribution of mean $1/\lambda$ and with probability $1 - \alpha$ from an exponential distribution with mean $1/\mu$. Service is exponential with mean $1/\mu$.

(a) Find $A^*(s)$ and $B^*(s)$.

(b) Find the expression for $\Psi_+(s)/\Psi_-(s)$ and show the pole-zero plot in the s-plane.

(c) Find $\Psi_+(s)$ and $\Psi_-(s)$.

(d) Find $\Phi_+(s)$ and $W(y)$.

SOLUTION

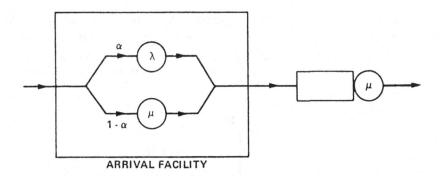

ARRIVAL FACILITY

We observe that this is an $H_2/M/1$ system. Also, $\bar{x} = 1/\mu$ and $\bar{t} = \alpha/\lambda + (1-\alpha)/\mu$ or $\bar{t} = 1/\mu + \alpha\left(1/\lambda - 1/\mu\right)$. Hence the system is stable if and only if

$$\alpha > 0 \quad \text{and} \quad 0 < \frac{\lambda}{\mu} < 1$$

(a) The interarrival time and service time pdf's are

$$a(t) = \alpha\lambda e^{-\lambda t} + (1-\alpha)\mu e^{-\mu t} \qquad t \geq 0$$

$$b(x) = \mu e^{-\mu x} \qquad x \geq 0$$

Thus

$$A^*(s) = \frac{\alpha\lambda}{s+\lambda} + \frac{(1-\alpha)\mu}{s+\mu}$$

$$B^*(s) = \frac{\mu}{s+\mu}$$

(b) We have

$$\frac{\Psi_+(s)}{\Psi_-(s)} = A^*(-s)B^*(s) - 1 = \left[\frac{\alpha\lambda}{\lambda - s} + \frac{(1-\alpha)\mu}{\mu - s}\right]\frac{\mu}{s+\mu} - 1$$

$$\frac{\Psi_+(s)}{\Psi_-(s)} = \frac{(-\alpha\lambda s + \mu\lambda - \mu s + \alpha\mu s)\mu - (\lambda - s)(\mu^2 - s^2)}{(\lambda - s)(\mu - s)(\mu + s)}$$

$$\frac{\Psi_+(s)}{\Psi_-(s)} = \frac{-s[s^2 - \lambda s - \alpha\mu(\mu - \lambda)]}{(\lambda - s)(\mu - s)(\mu + s)}$$

The roots of $s^2 - \lambda s - \alpha\mu(\mu - \lambda)$ are

$$s_1 \triangleq \frac{\lambda + \sqrt{\lambda^2 + 4\alpha\mu(\mu - \lambda)}}{2}$$

$$s_2 \triangleq \frac{\lambda - \sqrt{\lambda^2 + 4\alpha\mu(\mu - \lambda)}}{2}$$

$$\frac{\Psi_+(s)}{\Psi_-(s)} = \frac{-s(s - s_1)(s - s_2)}{(\lambda - s)(\mu - s)(\mu + s)}$$

To locate the relative positions of the poles and zeros, we note that for a stable system $(\alpha > 0, \lambda < \mu)$, then $s_1 > (\lambda + \sqrt{\lambda^2})/2 = \lambda$. Further, since also $\alpha \le 1$, then $s_1 \le [\lambda + \sqrt{\lambda^2 + 4\mu(\mu - \lambda)}]/2$ or $s_1 \le [\lambda + (2\mu - \lambda)]/2 = \mu$. So we have $\lambda < s_1 \le \mu$. Similarly, $\lambda - \mu \le s_2 < 0$. The pole-zero plot is as follows:

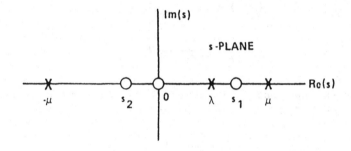

(c) Since $\Psi_+(s)$ must be analytic and zero-free in $\mathrm{Re}(s) > 0$ and since $\lim_{|s| \to \infty} \Psi_+(s)/s = 1$, we see that

$$\Psi_+(s) = s\left(\frac{s - s_2}{s + \mu}\right)$$

$$\Psi_-(s) = -\frac{(\lambda - s)(\mu - s)}{s - s_1}$$

(d) Equation (1.153) gives $\Phi_+(s) = K/\Psi_+(s)$. By Eq. (1.154)

$$K = \lim_{s \to 0} \frac{\Psi_+(s)}{s}$$

So

$$K = \lim_{s \to 0} \frac{s - s_2}{s + \mu} = \frac{-s_2}{\mu}$$

Thus

$$\Phi_+(s) = -\frac{s_2}{\mu}\frac{s + \mu}{s(s - s_2)}$$

Next we invert this transform using partial fraction expansion techniques to find

$$\Phi_+(s) = -\frac{s_2}{\mu}\left[\frac{-\mu/s_2}{s} + \frac{\mu + s_2}{s_2(s - s_2)}\right]$$

$$\Phi_+(s) = \frac{1}{s} - \frac{1 + s_2/\mu}{s - s_2}$$

Inverting we obtain

$$W(y) = 1 - \left(1 + \frac{s_2}{\mu}\right)e^{s_2 y} \qquad y \geq 0$$

[Note: This $H_2/M/1$ system is a G/M/1 system, where it is easily seen that $\sigma = 1 + s_2/\mu$ or $-s_2 = \mu(1 - \sigma)$.] □

PROBLEM 7.8

Consider a G/G/1 system in which

$$A^*(s) = \frac{2}{(s + 1)(s + 2)}$$

$$B^*(s) = \frac{1}{s + 1}$$

(a) Find the expression for $\Psi_+(s)/\Psi_-(s)$ and show the pole-zero plot in the s-plane.
(b) Use the method of spectrum factorization to find $\Psi_+(s)$ and $\Psi_-(s)$.
(c) Find $\Phi_+(s)$.
(d) Find $W(y)$.
(e) Find the average waiting time W.
(f) We solved for $W(y)$ by the method of spectrum factorization. Can you describe another way to find $W(y)$?

SOLUTION

(a) We have

$$\frac{\Psi_+(s)}{\Psi_-(s)} = A^*(-s)B^*(s) - 1 = \frac{2}{(1 - s)(2 - s)}\frac{1}{s + 1} - 1$$

$$= \frac{-s(s^2 - 2s - 1)}{(1 - s)(2 - s)(1 + s)}$$

$$= \frac{-s[s - (1 + \sqrt{2})][s - (1 - \sqrt{2})]}{(1 - s)(2 - s)(1 + s)}$$

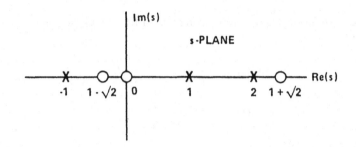

(b) Since $\Psi_+(s)$ must be analytic and zero-free in $\mathrm{Re}(s) > 0$ and since $\lim_{|s| \to +\infty} \Psi_+(s)/s = 1$, we see that

$$\Psi_+(s) = \frac{s[s - (1 - \sqrt{2})]}{s + 1}$$

$$\Psi_-(s) = -\frac{(1 - s)(2 - s)}{s - (1 + \sqrt{2})}$$

(pick $0 < D \le 1$).

(c) We have

$$\Phi_+(s) = \frac{K}{\Psi_+(s)}$$

where $K = \lim_{s \to 0} \Psi_+(s)/s$. Thus $K = \sqrt{2} - 1$ and so

$$\Phi_+(s) = (\sqrt{2} - 1)\frac{s + 1}{s[s - (1 - \sqrt{2})]}$$

or

$$\Phi_+(s) = \frac{1}{s} - \frac{2 - \sqrt{2}}{s - (1 - \sqrt{2})}$$

(d) Inverting this last we obtain

$$W(y) = 1 - (2 - \sqrt{2})e^{-(\sqrt{2} - 1)y} \qquad y \ge 0$$

(e) For W, we recognize that $W(y)$ is exponential in character. Indeed its density is

$$w(y) = (\sqrt{2} - 1)u_0(y) + (2 - \sqrt{2})(\sqrt{2} - 1)e^{-(\sqrt{2} - 1)y} \qquad y \ge 0$$

By inspection, the mean wait W is

$$W = \frac{2 - \sqrt{2}}{\sqrt{2} - 1} = \sqrt{2}$$

(f) Noting that this is a G/M/1 system, we may use the results from Section 1.8 to solve for $W(y)$. Thus, we first find σ satisfying $\sigma = A^*(\mu - \mu\sigma)$ (this gives $\sigma = 2 - \sqrt{2}$) and then

$$W(y) = 1 - \sigma e^{-\mu(1-\sigma)y} \qquad y \geq 0$$

by Eq. (1.126). □

PROBLEM 7.9

Consider the system M/G/1. Using the spectral solution method for Lindley's integral equation, find

(a) $\Psi_+(s)$. [HINT: Interpret $[1 - B^*(s)]/s\bar{x}$.]

(b) $\Psi_-(s)$.

(c) $s\Phi_+(s)$.

SOLUTION

(a) For M/G/1, $A^*(s) = \lambda/(s + \lambda)$. Thus

$$\frac{\Psi_+(s)}{\Psi_-(s)} = A^*(-s)B^*(s) - 1$$

$$= \frac{\lambda}{\lambda - s}B^*(s) - 1$$

$$= \frac{s - \lambda + \lambda B^*(s)}{\lambda - s}$$

Factoring the zero at $s = 0$ gives

$$\frac{\Psi_+(s)}{\Psi_-(s)} = \frac{s\left[1 - \dfrac{\lambda\bar{x}[1 - B^*(s)]}{s\bar{x}}\right]}{\lambda - s}$$

$$= \frac{s[1 - \rho\hat{B}^*(s)]}{\lambda - s}$$

where we recall from the footnote following Eq. (1.101) that

$$\hat{B}^*(s) \triangleq \frac{1 - B^*(s)}{s\bar{x}}$$

is the transform of the residual life pdf for the service time. We know that the density function of a non-negative random variable has a Laplace transfor..., say, $H^*(s)$, which, for $\mathrm{Re}(s) \geq 0$, is analytic and satisfies $|H^*(s)| \leq 1$.

Applying this to $\hat{B}^*(s)$ we find

$$|\rho\hat{B}^*(s)| \le \rho < 1 \qquad \text{for } \text{Re}(s) \ge 0$$

Thus $1 - \rho\hat{B}^*(s)$ has no zeros (and no poles) in $\text{Re}(s) \ge 0$, and so this function must be part of $\Psi_+(s)$. Since $\Psi_+(s)$ must be analytic and zero-free in $\text{Re}(s) > 0$ and $\lim_{|s|\to\infty} \Psi_+(s)/s = 1$, we have

$$\Psi_+(s) = s[1 - \rho\hat{B}^*(s)] = s - \lambda + \lambda B^*(s)$$

[Note that $\lim_{|s|\to\infty} \hat{B}^*(s) = 0$ for $\text{Re}(s) > 0$, so $\lim_{|s|\to\infty} \Psi_+(s)/s = 1$.]
(b) Clearly,

$$\Psi_-(s) = \lambda - s$$

(pick $0 < D \le \lambda$).
(c) We have $\Phi_+(s) = K/\Psi_+(s)$, where $K = \lim_{s\to 0} \Psi_+(s)/s$. Thus

$$K = \lim_{s\to 0}[1 - \rho\hat{B}^*(s)] = 1 - \rho$$

and

$$\Phi_+(s) = \frac{1 - \rho}{s - \lambda + \lambda B^*(s)}$$

Finally

$$s\Phi_+(s) = \frac{s(1 - \rho)}{s - \lambda + \lambda B^*(s)} = W^*(s)$$

This is Eq. (1.105), the P-K transform equation for the waiting-time distribution.   □

PROBLEM 7.10

Consider the queue $E_q/E_r/1$.

(a) Show that

$$\frac{\Psi_+(s)}{\Psi_-(s)} = \frac{F(s)}{1 - F(s)}$$

where $F(s) = 1 - (1 - s/\lambda q)^q(1 + s/\mu r)^r$.
(b) For $\rho < 1$, show that $F(s)$ has one zero at the origin, zeros $s_1, s_2, \ldots, s_r$ in $\text{Re}(s) < 0$, and zeros $s_{r+1}, s_{r+2}, \ldots, s_{r+q-1}$ in $\text{Re}(s) > 0$.
(c) Express $\Psi_+(s)$ and $\Psi_-(s)$ in terms of s_i.
(d) Express $W^*(s)$ in terms of s_i ($i = 1, 2, \ldots, r + q - 1$).

SOLUTION

(a) For $E_q/E_r/1$, we have $A^*(s) = [\lambda q/(s + \lambda q)]^q$ and $B^*(s) = [\mu r/(s + \mu r)]^r$. (Note that $\bar{t} = 1/\lambda$, $\bar{x} = 1/\mu$ and so $\rho = \lambda/\mu$.)

$$
\frac{\Psi_+(s)}{\Psi_-(s)} = A^*(-s)B^*(s) - 1 = \left(\frac{\lambda q}{\lambda q - s}\right)^q \left(\frac{\mu r}{\mu r + s}\right)^r - 1
$$

$$
= \frac{1}{\left(1 - \dfrac{s}{\lambda q}\right)^q \left(1 + \dfrac{s}{\mu r}\right)^r} - 1
$$

$$
= \frac{1 - \left(1 - \dfrac{s}{\lambda q}\right)^q \left(1 + \dfrac{s}{\mu r}\right)^r}{\left(1 - \dfrac{s}{\lambda q}\right)^q \left(1 + \dfrac{s}{\mu r}\right)^r}
$$

Thus

$$
\frac{\Psi_+(s)}{\Psi_-(s)} = \frac{F(s)}{1 - F(s)}
$$

where

$$
F(s) = 1 - \left(1 - \frac{s}{\lambda q}\right)^q \left(1 + \frac{s}{\mu r}\right)^r
$$

(b) We assume $\rho < 1$, or $\lambda < \mu$. We wish to apply Rouché's theorem to $F(s)$. Thus we define the functions $f(s)$ and $g(s)$ as follows:

$$
f(s) = -\left(1 - \frac{s}{\lambda q}\right)^q \left(1 + \frac{s}{\mu r}\right)^r \quad \text{and} \quad g(s) = 1
$$

Therefore $F(s) = f(s) + g(s)$. We first consider two circles: let K_1 be the circle of radius μr centered at the point $s \overset{\Delta}{=} (x, y) = (-\mu r, 0)$ and let K_2 be the circle of radius λq centered at the point $s \overset{\Delta}{=} (x, y) = (\lambda q, 0)$.

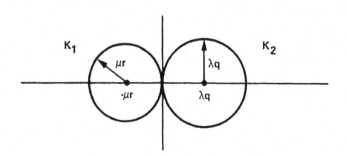

Any point $s = (x, y)$ on K_1 must satisfy $(x + \mu r)^2 + y^2 = (\mu r)^2$. Thus we have

$$|1 + s/\mu r|^2 = \frac{(x + \mu r)^2 + y^2}{(\mu r)^2} = 1$$

on K_1. Hence outside K_1, $|1 + s/\mu r|^2 > 1$. Similarly, $|1 - s/\lambda q|^2 = 1$ on K_2 and $|1 - s/\lambda q|^2 > 1$ outside K_2. Thus $|f(s)| > 1 = |g(s)|$ for any point s that is outside one of the circles and on or outside the other. Note that $s = 0$ does not satisfy this condition, and to use Rouché's theorem we will have to detour around the origin. To this end, let $K_3(\varepsilon)$ be a semicircle of radius $\varepsilon > 0$ about the origin in the left-hand plane, that is, in $\mathrm{Re}(s) < 0$. On $K_3(\varepsilon)$, $s = (x, y)$ must satisfy $x^2 + y^2 = \varepsilon^2$, $x = \varepsilon \cos \theta$, $y = \varepsilon \sin \theta$, where $\pi/2 < \theta < 3\pi/2$ (i.e., $\cos \theta < 0$).

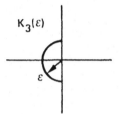

On $K_3(\varepsilon)$ we have

$$|1 - s/\lambda q|^2 = \frac{(\lambda q - x)^2 + y^2}{(\lambda q)^2} = \frac{(\lambda q)^2 - 2\lambda q x + \varepsilon^2}{(\lambda q)^2}$$

Since $\varepsilon > 0$, then we have

$$|1 - s/\lambda q|^2 > \frac{(\lambda q)^2 - 2\lambda q x}{(\lambda q)^2} = 1 - \frac{2\varepsilon \cos \theta}{\lambda q}$$

Similarly, on $K_3(\varepsilon)$

$$|1 + s/\mu r|^2 > 1 + \frac{2\varepsilon \cos \theta}{\mu r}$$

Thus, on $K_3(\varepsilon)$, we have

$$
\begin{aligned}
|f(s)|^2 &= \left| -\left(1 - \frac{s}{\lambda q}\right)^q \left(1 + \frac{s}{\mu r}\right)^r \right|^2 \\
&> \left(1 - \frac{2\varepsilon \cos \theta}{\lambda q}\right)^q \left(1 + \frac{2\varepsilon \cos \theta}{\mu r}\right)^r \\
&= \left(1 - \frac{2\varepsilon \cos \theta}{\lambda} + o(\varepsilon)\right) \left(1 + \frac{2\varepsilon \cos \theta}{\mu} + o(\varepsilon)\right)
\end{aligned}
$$

and so

$$|f(s)|^2 > 1 - \frac{2\varepsilon \cos \theta}{\lambda}(1 - \rho) + o(\varepsilon)$$

Since $\cos \theta < 0$ and $\rho < 1$, then $-[(2\varepsilon \cos \theta)/\lambda](1 - \rho) > 0$. Thus from the definition of $o(\varepsilon)$, there exists $\delta > 0$ such that, for $0 < \varepsilon < \delta$, $|f(s)| > 1 = |g(s)|$ on $K_3(\varepsilon)$. Now we apply Rouché's theorem as follows. Let $C_1(\varepsilon)$ be the closed contour consisting of part of K_1 plus the detour around the origin in $\text{Re}(s) < 0$ $(0 < \varepsilon < \delta)$.

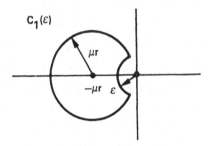

Then $|f(s)| > 1 = |g(s)|$ on $C_1(\varepsilon)$. Now $f(s)$ has r roots inside $C_1(\varepsilon)$, all at $s = (-\mu r, 0)$. Thus $F(s) = f(s) + g(s)$ has r roots inside $C_1(\varepsilon)$. Letting $\varepsilon \to 0$ we see that $F(s)$ has r zeros $s_1, \dots, s_r$ inside K_1, the circle of radius μr about $s = (-\mu r, 0)$. Next let $C_2(\varepsilon)$ be the closed contour consisting of part of K_2 together with part of a circle about the origin of radius ε $[C_2(\varepsilon)$ contains the origin].

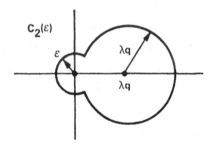

As noted above, for any point s on $C_2(\varepsilon)$ outside K_1 [which includes all points on $C_2(\varepsilon)$ in $\text{Re}(s) \geq 0$] we have $|f(s)| > 1 = |g(s)|$. For points s on $C_2(\varepsilon)$ in $\text{Re}(s) < 0$ we use the same argument as before. Thus $|f(s)| > 1 = |g(s)|$ on $C_2(\varepsilon)$. Now $f(s)$ has q zeros inside $C_2(\varepsilon)$ [all at $s = (\lambda q, 0)$], so $F(s)$ has q zeros inside $C_2(\varepsilon)$. By examining $F(s)$ (recall that $\rho < 1$) we see that exactly one of these zeros is at $s = 0$. Letting $\varepsilon \to 0$ we see that $F(s)$ has $q - 1$ zeros $s_{r+1}, \dots, s_{r+q-1}$ inside the circle K_2. In summary: $F(s)$ has r zeros $s_1, \dots, s_r$ inside the circle of radius μr centered at $s = (-\mu r, 0)$, $F(s)$ has one zero at the origin, and $F(s)$ has $q - 1$ zeros $s_{r+1}, \dots, s_{r+q-1}$ inside the circle of radius λq centered at $s = (\lambda q, 0)$. This (stronger) statement implies part (b).

(c) From part (b), we can write

$$F(s) = 1 - \left(1 - \frac{s}{\lambda q}\right)^q \left(1 + \frac{s}{\mu r}\right)^r$$

$$= -\left(\frac{-1}{\lambda q}\right)^q \left(\frac{1}{\mu r}\right)^r s(s - s_1) \cdots (s - s_{r+q-1})$$

So

$$\frac{\Psi_+(s)}{\Psi_-(s)} = \frac{F(s)}{1 - F(s)} = \frac{F(s)}{\left(1 - \dfrac{s}{\lambda q}\right)^q \left(1 + \dfrac{s}{\mu r}\right)^r}$$

$$= -\frac{s(s - s_1) \cdots (s - s_{r+q-1})}{(s - \lambda q)^q (s + \mu r)^r}$$

We know that $\Psi_+(s)$ must absorb all the poles and zeros in the left-hand plane, thus

$$\Psi_+(s) = \frac{s(s - s_1) \cdots (s - s_r)}{(s + \mu r)^r}$$

$$\Psi_-(s) = -\frac{(s - \lambda q)^q}{(s - s_{r+1}) \cdots (s - s_{r+q-1})}$$

or

$$\Psi_+(s) = \frac{s \prod_{i=1}^{r}(s - s_i)}{(s + \mu r)^r}$$

$$\Psi_-(s) = -\frac{(s - \lambda q)^q}{\prod_{i=r+1}^{r+q-1}(s - s_i)}$$

(d) We have $\Phi_+(s) = K/\Psi_+(s)$, where $K = \lim_{s \to 0} \Psi_+(s)/s$. Thus

$$K = \frac{\prod_{i=1}^{r}(-s_i)}{(\mu r)^r}$$

and so

$$\Phi_+(s) = \frac{\prod_{i=1}^{r}(-s_i)}{(\mu r)^r} \cdot \frac{(s + \mu r)^r}{s \prod_{i=1}^{r}(s - s_i)}$$

$$W^*(s) = s\Phi_+(s) = \frac{\left(1 + \dfrac{s}{\mu r}\right)^r}{\prod_{i=1}^{r}\left(1 - \dfrac{s}{s_i}\right)} \qquad \Box$$

PROBLEM 7.11

For G/M/1, it can be shown that (see [KLEI 75], pages 292–293)

$$W(y) = 1 - \left(1 - \frac{s_1}{\mu}\right) e^{-s_1 y} \qquad y \geq 0$$

where $-s_1$ is the only zero of $s + \mu - \mu A^*(-s)$ satisfying $\text{Re}(s) < 0$. Show that this is equivalent to Eq. (1.126).

SOLUTION

By definition, $-s_1 + \mu = \mu A^*(s_1)$. Define $\sigma = 1 - s_1/\mu$. Therefore $s_1 = \mu - \mu\sigma$. So we have $\mu\sigma = \mu A^*(\mu - \mu\sigma)$ or $\sigma = A^*(\mu - \mu\sigma)$. The stated result then becomes

$$W(y) = 1 - \sigma e^{-\mu(1-\sigma)y} \qquad y \geq 0$$

which is Eq. (1.126).   □

PROBLEM 7.12

Consider a D/D/1 queue with $\rho < 1$. Assume $w_0 = 4\bar{t}(1 - \rho)$.

(a) Calculate $w_n(y)$ using the procedure defined by Eq. (1.145) for $n = 0, 1, 2, \ldots$.
(b) Show that the known solution for

$$w(y) = \lim_{n \to \infty} w_n(y)$$

satisfies Eq. (1.146).

SOLUTION

(a) Equation (1.145) states

$$w_{n+1}(y) = \pi(c(y) \circledast w_n(y)) \qquad n = 1, 2, \ldots$$

For D/D/1 we know

$$a(y) = u_0(y - \bar{t})$$
$$b(y) = u_0(y - \bar{x})$$

Then, using the convolution notation $\circledast$, Eq. (1.142) gives

$$c(y) = a(-y) \circledast b(y)$$

So

$$c(y) = u_0(y + \bar{t} - \bar{x}) = u_0(y + \bar{t}(1 - \rho))$$

Thus

$$c(y) \circledast w_n(y) = \int_{-\infty}^{y} w_n(y - t)c(t)\,dt$$
$$= w_n(y + \bar{t}(1 - \rho))$$

So

$$w_{n+1}(y) = \pi(w_n(y + \bar{t}(1 - \rho)))$$

Using this relationship, w_n may be calculated for $n = 0, 1, 2, \ldots$. By assumption

$$w_0 = u_0(y - 4\bar{t}(1 - \rho))$$

Also

$$w_1(y) = \pi(w_0(y + \bar{t}(1 - \rho))) = \pi(u_0(y - 3\bar{t}(1 - \rho)))$$

Since the sweep operator π has no "negative" probability to sweep up, we find

$$w_1(y) = u_0(y - 3\bar{t}(1 - \rho))$$

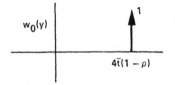
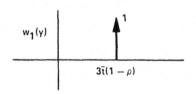

Also

$$w_2(y) = \pi(w_1(y + \bar{t}(1 - \rho))) = \pi(u_0(y - 2\bar{t}(1 - \rho)))$$
$$w_2(y) = u_0(y - 2\bar{t}(1 - \rho))$$

Similarly,

$$w_3(y) = u_0(y - \bar{t}(1 - \rho))$$

and

$$w_4(y) = u_0(y)$$

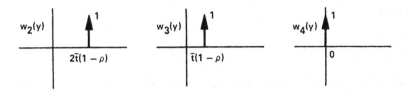

In fact,

$$w_n(y) = u_0(y) \qquad \text{for } n \geq 4$$

(due to sweeping)

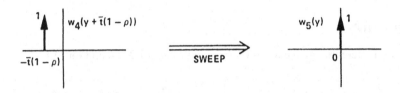

(b) We have

$$w(y) = \lim_{n \to \infty} w_n(y) = u_0(y)$$

Then

$$c(y) \circledast w(y) = \int_{-\infty}^{y} w(y - t)c(t)\, dt$$

$$= \int_{-\infty}^{y} w(y - t)u_0(t + \bar{t}(1 - \rho))\, dt$$

$$= w(y + \bar{t}(1 - \rho)) = u_0(y + \bar{t}(1 - \rho))$$

So

$$\pi(c(y) \circledast w(y)) = \pi(u_0(y + \bar{t}(1 - \rho)))$$

$$= u_0(y)$$

[the unit impulse at $y = -\bar{t}(1 - \rho)$ is swept up to the origin]. Thus

$$\pi(c(y) \circledast w(y)) = w(y)$$

and Eq. (1.146) is satisfied.   □

PROBLEM 7.13

Consider an M/M/1 queue with $\rho < 1$. Assume $w_0 = 0$.

(a) Calculate $w_1(y)$ using the procedure defined by Eq. (1.145).

(b) Repeat for $w_2(y)$.

(c) Show that our known solution for

$$w(y) = \lim_{n \to \infty} w_n(y)$$

satisfies Eq. (1.146).

(d) Compare $w_2(y)$ with $w(y)$.

SOLUTION

(a) Differentiating the result for $C(y)$ in Problem 7.2, we find for M/M/1

$$c(y) = \begin{cases} \dfrac{\lambda\mu}{\lambda + \mu} e^{-\mu y} & y \geq 0 \\[2mm] \dfrac{\lambda\mu}{\lambda + \mu} e^{\lambda y} & y < 0 \end{cases}$$

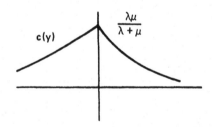

$w_0 = 0$ implies $w_0(y) = u_0(y)$ $(y \geq 0)$. By Eq. (1.145) we have

$$w_1(y) = \pi(c(y) \circledast w_0(y))$$

$$w_1(y) = \pi(c(y) \circledast u_0(y)) = \pi(c(y))$$

Thus

$$w_1(y) = u_0(y) \int_{-\infty}^{0} \frac{\lambda\mu}{\lambda + \mu} e^{\lambda y}\, dy + \frac{\lambda\mu}{\lambda + \mu} e^{-\mu y}$$

$$w_1(y) = \frac{\mu}{\lambda + \mu} u_0(y) + \frac{\lambda\mu}{\lambda + \mu} e^{-\mu y} \qquad y \geq 0$$

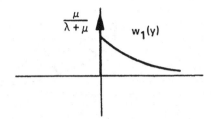

(b) We have

$$w_2(y) = \pi(c(y) \circledast w_1(y))$$

We first determine $c(y) \circledast w_1(y)$.

$$c(y) \circledast w_1(y) = \int_{-\infty}^{\infty} c(y - t)w_1(t)\, dt = \int_{0}^{\infty} c(y - t)w_1(t)\, dt$$

Case (1): $y \leq 0$

Here $y - t \leq 0$ for all $0 \leq t < \infty$.

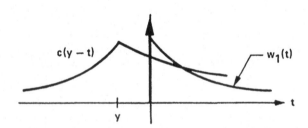

Thus

$$c(y) \circledast w_1(y) = \int_{t}^{\infty} \frac{\lambda\mu}{\lambda + \mu} e^{\lambda(y-t)} \left[\frac{\mu}{\lambda + \mu} u_0(t) + \frac{\lambda\mu}{\lambda + \mu} e^{-\mu t} \right] dt$$

$$= \frac{\lambda\mu^2}{(\lambda + \mu)^2} e^{\lambda y} + \left(\frac{\lambda\mu}{\lambda + \mu} \right)^2 e^{\lambda y} \int_{0}^{\infty} e^{-(\lambda + \mu)t}\, dt$$

$$= \frac{\lambda\mu^2}{(\lambda + \mu)^2} e^{\lambda y} \left(1 + \frac{\lambda}{\lambda + \mu} \right)$$

Case (2): $y > 0$

We consider the intervals $y - t \geq 0$ and $y - t \leq 0$, that is, $0 \leq t \leq y$ and $y \leq t < \infty$:

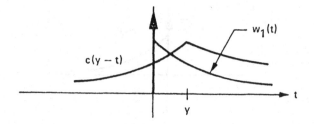

We have

$$c(y) \circledast w_1(y) = \int_0^y \frac{\lambda\mu}{\lambda + \mu} e^{-\mu(y-t)} \left[\frac{\mu}{\lambda + \mu} u_0(t) + \frac{\lambda\mu}{\lambda + \mu} e^{-\mu t} \right] dt$$

$$+ \int_y^x \frac{\lambda\mu}{\lambda + \mu} e^{\lambda(y-t)} \left[\frac{\mu}{\lambda + \mu} u_0(t) + \frac{\lambda\mu}{\lambda + \mu} e^{-\mu t} \right] dt$$

or

$$c(y) \circledast w_1(y) = \frac{\lambda\mu^2}{(\lambda + \mu)^2} e^{-\mu y} + \left(\frac{\lambda\mu}{\lambda + \mu} \right)^2 e^{-\mu y} \int_0^y dt$$

$$+ \left(\frac{\lambda\mu}{\lambda + \mu} \right)^2 e^{\lambda y} \int_y^x e^{-(\lambda + \mu)t} dt$$

Thus

$$c(y) \circledast w_1(y) = \frac{\lambda\mu^2}{(\lambda + \mu)^2} e^{-\mu y}(1 + \lambda y) + \frac{(\lambda\mu)^2}{(\lambda + \mu)^3} e^{\lambda y} [e^{-(\lambda + \mu)y}]$$

$$= \frac{\lambda\mu^2}{(\lambda + \mu)^2} e^{-\mu y} \left(1 + \lambda y + \frac{\lambda}{\lambda + \mu} \right)$$

To find the probability in the negative half-line, we calculate

$$\int_{-x}^0 c(y) \circledast w_1(y) \, dy = \int_{-x}^0 \frac{\lambda\mu^2}{(\lambda + \mu)^2} e^{\lambda y} \left(1 + \frac{\lambda}{\lambda + \mu} \right) dy$$

$$= \frac{\lambda\mu^2}{(\lambda + \mu)^2} \left(1 + \frac{\lambda}{\lambda + \mu} \right) \frac{e^{\lambda y}}{\lambda} \Big|_{-x}^0$$

$$= \left(\frac{\mu}{\lambda + \mu} \right)^2 \left(1 + \frac{\lambda}{\lambda + \mu} \right) = \frac{\mu^2(2\lambda + \mu)}{(\lambda + \mu)^3}$$

Thus

$$w_2(y) = \pi(c(y) \circledast w_1(y))$$

$$w_2(y) = \frac{\mu^2(2\lambda + \mu)}{(\lambda + \mu)^3} u_0(y) + \frac{\lambda\mu^2}{(\lambda + \mu)^2} e^{-\mu y} \left(1 + \lambda y + \frac{\lambda}{\lambda + \mu} \right) \qquad y \geq 0$$

(c) The known solution for M/M/1 as given in Eq. (1.70) is

$$w(y) = (1 - \rho)u_0(y) + \lambda(1 - \rho)e^{-\mu(1-\rho)y} \qquad y \geq 0$$

We first find $c(y) \circledast w(y)$.

$$c(y) \circledast w(y) = \int_{-\infty}^{\infty} c(y - t)w(t)\, dt = \int_0^{\infty} c(y - t)w(t)\, dt$$

For $y \leq 0$ then $y - t \leq 0$, so

$$c(y) \circledast w(y) = \int_0^{\infty} \frac{\lambda\mu}{\lambda + \mu} e^{\lambda(y-t)} \left[(1 - \rho)u_0(t) + \lambda(1 - \rho)e^{-\mu(1-\rho)t}\right] dt$$

$$= \frac{\lambda\mu}{\lambda + \mu} e^{\lambda y}(1 - \rho) + \frac{\lambda\mu}{\lambda + \mu} e^{\lambda y}\lambda(1 - \rho) \int_0^{\infty} e^{-\mu t}\, dt$$

$$= \frac{\lambda\mu}{\lambda + \mu} e^{\lambda y}(1 - \rho)\left(1 + \frac{\lambda}{\mu}\right) = \lambda(1 - \rho)e^{\lambda y}$$

Thus the probability π will sweep up is

$$\int_{-\infty}^{0} c(y) \circledast w(y)\, dy = \int_{-\infty}^{0} \lambda(1 - \rho)e^{\lambda y} = 1 - \rho$$

For $y \geq 0$ we consider $0 \leq t \leq y$ and $y \leq t < \infty$. Then

$$c(y) \circledast w(y) = \int_0^{y} \frac{\lambda\mu}{\lambda + \mu} e^{-\mu(y-t)} \left[(1 - \rho)u_0(t) + \lambda(1 - \rho)e^{-\mu(1-\rho)t}\right] dt$$

$$+ \int_{y}^{\infty} \frac{\lambda\mu}{\lambda + \mu} e^{\lambda(y-t)} \left[(1 - \rho)u_0(t) + \lambda(1 - \rho)e^{-\mu(1-\rho)t}\right] dt$$

or

$$c(y) \circledast w(y) = \frac{\lambda\mu}{\lambda + \mu} e^{-\mu y}(1 - \rho) + \frac{\lambda\mu}{\lambda + \mu} e^{-\mu y}\lambda(1 - \rho) \int_0^{y} e^{\lambda t}\, dt$$

$$+ \frac{\lambda\mu}{\lambda + \mu} e^{\lambda y}\lambda(1 - \rho) \int_{y}^{\infty} e^{-\mu t}\, dt$$

$$= \frac{\lambda\mu}{\lambda + \mu} e^{-\mu y}(1 - \rho)e^{\lambda y} + \frac{\lambda\mu}{\lambda + \mu} e^{\lambda y}\lambda(1 - \rho)\frac{e^{-\mu y}}{\mu}$$

$$= \frac{\lambda\mu}{\lambda + \mu} e^{-(\mu - \lambda)y}(1 - \rho)\left(1 + \frac{\lambda}{\mu}\right)$$

$$= \lambda(1 - \rho)e^{-\mu(1-\rho)y}$$

Thus

$$\pi(c(y) \circledast w(y)) = (1 - \rho)u_0(y) + \lambda(1 - \rho)e^{-\mu(1-\rho)y} \qquad y \geq 0$$

and

$$w(y) = \pi(c(y) \circledast w(y))$$

Hence $w(y)$ satisfies Eq. (1.146).

(d) We have

$$w(y) = (1 - \rho)u_0(y) + \lambda(1 - \rho)e^{-\mu(1-\rho)y} \qquad y \geq 0$$

$$w_2(y) = \frac{\mu^2(2\lambda + \mu)}{(\lambda + \mu)^3}u_0(y) + \frac{\lambda\mu^2}{(\lambda + \mu)^2}e^{-\mu y}\left(1 + \lambda y + \frac{\lambda}{\lambda + \mu}\right)$$

$$y \geq 0$$

$$P[w_2 = 0] = \frac{\mu^2(2\lambda + \mu)}{(\lambda + \mu)^3} = \frac{1 + 2\rho}{(1 + \rho)^3}$$

$$> \frac{(1 + 2\rho + \rho^2)(1 - \rho^2)}{(1 + \rho)^3} = 1 - \rho$$

or

$$P[w_2 = 0] > P[\tilde{w} = 0]$$

So $w_2(y)$ has a larger impulse at the origin than $w(y)$. Also $w(y)$ decays as $e^{-\mu(1-\rho)y} = e^{-\mu y}e^{\lambda y}$, whereas $w_2(y)$ decays as $ye^{-\mu y}$. Thus $w_2(y)$ decays more rapidly. $\qquad\square$

PROBLEM 7.14

By first cubing Eq. (1.149) and then forming expectations, express σ_w^2 (the variance of the waiting time) in terms of the first three moments of $\tilde{t}$, $\tilde{x}$, and I.

SOLUTION

Cubing Eq. (1.149) $(w_{n+1} - y_n = w_n + u_n)$ gives

$$w_{n+1}^3 - 3w_{n+1}^2 y_n + 3w_{n+1}y_n^2 - y_n^3 = w_n^3 + 3w_n^2 u_n + 3w_n u_n^2 + u_n^3$$

Recalling that $w_{n+1}y_n = 0$, taking expectations and letting $n \to \infty$, we get

$$\overline{w^3} - \overline{y^3} = \overline{w^3} + 3\overline{w^2}\,\overline{u} + 3\overline{w}\,\overline{u^2} + \overline{u^3}$$

Since $\tilde{w}$ and $\tilde{u}$ are independent we find

$$\overline{w^2} = -\frac{\overline{y^3} + \overline{u^3}}{3\overline{u}} - \frac{\overline{w}\,\overline{u^2}}{\overline{u}}$$

Thus

$$\sigma_w^2 = \overline{w^2} - (\overline{w})^2 = -\frac{\overline{y^3} + \overline{u^3}}{3\overline{u}} - \frac{\overline{w}\,\overline{u^2}}{\overline{u}} - (\overline{w})^2$$

Using Eq. (1.158), which states $\overline{w} = -\overline{u^2}/2\overline{u} - \overline{y^2}/2\overline{y}$, we get

$$\sigma_w^2 = -\frac{\overline{y^3} + \overline{u^3}}{3\overline{u}} + \frac{\overline{u^2}}{\overline{u}}\left(\frac{\overline{u^2}}{2\overline{u}} + \frac{\overline{y^2}}{2\overline{y}}\right) - \left(\frac{\overline{u^2}}{2\overline{u}} + \frac{\overline{y^2}}{2\overline{y}}\right)^2$$

$$= -\frac{\overline{y^3} + \overline{u^3}}{3\overline{u}} + \left(\frac{\overline{u^2}}{2\overline{u}} + \frac{\overline{y^2}}{2\overline{y}}\right)\left(\frac{\overline{u^2}}{2\overline{u}} - \frac{\overline{y^2}}{2\overline{y}}\right)$$

It can be shown (see [KLEI 75], pages 305–306) that $\overline{y^k} = a_0 \overline{t^k}$. Using this for $k = 1, 2$ and the fact [from Eq. (1.150)] that $\overline{y} = -\overline{u}$, we have

$$\sigma_w^2 = \frac{\overline{t^3}}{3\overline{t}} - \frac{\overline{u^3}}{3\overline{u}} + \left(\frac{\overline{u^2}}{2\overline{u}} + \frac{\overline{t^2}}{2\overline{t}}\right)\left(\frac{\overline{u^2}}{2\overline{u}} - \frac{\overline{t^2}}{2\overline{t}}\right)$$

Noting that $\tilde{u} = \tilde{x} - \tilde{t}$, we finally obtain

$$\sigma_w^2 = -\frac{\overline{x^3} - 3\overline{x^2}\,\overline{t} + 3\overline{x}\,\overline{t^2} - \overline{t^3}}{3(\overline{x} - \overline{t})} + \frac{1}{4}\left[\frac{\overline{x^2} - 2\overline{x}\,\overline{t} + \overline{t^2}}{\overline{x} - \overline{t}}\right]^2 + \frac{\overline{t^3}}{3\overline{t}} - \left(\frac{\overline{t^2}}{2\overline{t}}\right)^2$$

or

$$\sigma_w^2 = \frac{\overline{x^3} - \overline{t^3}}{3\overline{t}(1 - \rho)} + \frac{\rho\overline{t^2} - \overline{x^2}}{1 - \rho} + \left[\frac{\sigma_a^2 + \sigma_b^2 + [\overline{t}(1 - \rho)]^2}{2\overline{t}(1 - \rho)}\right]^2 + \frac{\overline{t^3}}{3\overline{t}} - \left(\frac{\overline{t^2}}{2\overline{t}}\right)^2 \qquad \Box$$

PROBLEM 7.15

Show that $P[\tilde{w} = 0] = 1 - \sigma$ from Eq. (1.161) by finding the constant term in a power-series expansion of $W^*(s)$.

SOLUTION

From Eq. (1.161) we see that

$$W^*(s) = (1 - \sigma)\sum_{j=0}^{\infty}[\sigma \hat{I}^*(s)]^j$$

Inverting [as we did in obtaining Eq. (1.108) from Eq. (1.107)] we have

$$w(y) = (1 - \sigma)\sum_{j=0}^{\infty}\sigma^j \hat{i}_{(j)}(y)$$

where the density function $\hat{i}_{(j)}(y)$ is the j-fold convolution of $\hat{i}(y)$ with itself. Thus

$$w(y) = (1 - \sigma)u_0(y) + (1 - \sigma)\sum_{j=1}^{\infty} \sigma^j \hat{i}_{(j)}(y)$$

In order to find $P[\tilde{w} = 0]$, we must determine for each $j = 1, 2, \ldots$ whether $\hat{i}_{(j)}(y)$ contains an impulse at the origin $(y = 0)$. By definition, any idle time must be greater than zero, and so we see that $P[\hat{I} = 0] = 0$. Thus $\hat{i}_{(j)}(y)$ cannot contain an impulse at the origin. Finally,

$$P[\tilde{w} = 0] = \int_0^{0^-} w(y)\,dy = 1 - \sigma + (1 - \sigma)\sum_{j=1}^{\infty}\sigma^j \int_{0-}^{0^-}\hat{i}_{(j)}(y)\,dy$$

or

$$P[\tilde{w} = 0] = 1 - \sigma \qquad\qquad \square$$

PROBLEM 7.16

Consider a G/G/1 system.

(a) Express $\hat{I}^*(s)$ in terms of $I^*(s)$, the transform of the pdf of idle time in the given system.

(b) Using (a) find $\hat{I}^*(s)$ when the original system is the ordinary M/M/1.

(c) Consider a G/M/1 queue.

   (i) For $\rho < 1$ (stable queue), use Eq. (1.160) to show directly that

   $$I^*(s) = \frac{A^*(s) - \sigma}{-\dfrac{s}{\mu} + 1 - \sigma}$$

   (ii) For $\rho > 1$ (unstable queue), use (a) to show that

   $$I^*(s) = \frac{1 - A^*(s)}{s\bar{t}}$$

(d) Since either the original or the dual queue must be unstable (except for D/D/1), discuss the existence of the transform of the idle-time pdf for the unstable queue.

SOLUTION

(a) Equation (1.160) and Eq. (1.161) give expressions for $W^*(s)$. Equating these two yields

$$\frac{a_0[1 - I^*(-s)]}{1 - C^*(s)} = W^*(s) = \frac{1 - \sigma}{1 - \sigma \hat{I}^*(s)}$$

Solving for $\hat{I}^*(s)$ we find

$$\hat{I}^*(s) = \frac{1}{\sigma}\left[1 - \frac{(1 - \sigma)(1 - C^*(s))}{a_0(1 - I^*(-s))}\right]$$

Here $1 - \sigma = P[\tilde{w} = 0]$ and $a_0 = P[\tilde{y} > 0]$. Note that $1 - \sigma = a_0$ if there is zero probability that an arrival and a departure occur at the same instant of time (e.g., when either the interarrival or service distribution is continuous). In this latter case we obtain the simplified expression

$$\hat{I}^*(s) = \frac{1}{\sigma}\left[1 - \frac{1 - C^*(s)}{1 - I^*(-s)}\right]$$

(b) For M/M/1, $1 - \sigma = a_0$ as noted above. Also, $\sigma = \rho$, $I^*(-s) = \lambda/(\lambda - s)$, and $C^*(s) = A^*(-s)B^*(s) = \lambda\mu/[(\mu + s)(\lambda - s)]$. Using part (a) we find

$$\hat{I}^*(s) = \frac{1}{\rho}\left[1 - \frac{1 - \dfrac{\lambda\mu}{(\mu + s)(\lambda - s)}}{1 - \dfrac{\lambda}{\lambda - s}}\right]$$

$$= \frac{1}{\rho}\left[1 - \frac{(\mu + s)(\lambda - s) - \lambda\mu}{(\mu + s)(-s)}\right]$$

$$= \frac{1}{\rho}\left[\frac{(\mu + s)(-\lambda) + \lambda\mu}{(\mu + s)(-s)}\right] = \frac{1}{\rho}\left(\frac{\lambda}{s + \mu}\right)$$

or

$$\hat{I}^*(s) = \frac{\mu}{s + \mu}$$

Thus the (unstable) dual queue has an exponential idle period distribution, which agrees with what we know since it is also M/M/1.

(c) (i) For a stable G/M/1 queue, from Eq. (1.126) we obtain

$$W^*(s) = \frac{(1 - \sigma)(s + \mu)}{s + \mu(1 - \sigma)}$$

Also $a_0 = 1 - \sigma$ and $C^*(s) = A^*(-s)\mu/(s + \mu)$. Thus Eq. (1.160) gives

$$\frac{(1 - \sigma)(s + \mu)}{s + \mu(1 - \sigma)} = \frac{(1 - \sigma)[1 - I^*(-s)]}{1 - A^*(-s)\dfrac{\mu}{s + \mu}}$$

or

$$1 - I^*(-s) = \frac{s + \mu - \mu A^*(-s)}{s + \mu - \mu\sigma}$$

Thus

$$I^*(-s) = \frac{\mu A^*(-s) - \mu\sigma}{s + \mu - \mu\sigma}$$

and so

$$I^*(s) = \frac{A^*(s) - \sigma}{-\dfrac{s}{\mu} + 1 - \sigma}$$

(ii) Here we must use caution since Eq. (1.161), and thus the result of part (a), requires that the original queue be stable (which does not hold for the G/M/1 system we are now considering). Thus we regard our unstable G/M/1 queue with parameters $A^*(s)$, μ as the dual of a stable M/G/1 queue with arrival rate μ and service time transform $A^*(s)$. For this stable M/G/1 system, $\sigma = \rho = \mu\bar{t}$. $I^*(s) = \mu/(s + \mu)$, and $C^*(s) = [\mu/(\mu - s)]A^*(s)$ (also note that $1 - \sigma = a_0$). We now use the result of part (a) to solve for $\hat{I}^*(s)$, which is, in fact, the idle time transform for our unstable G/M/1 queue. Thus

$$\hat{I}^*(s) = \frac{1}{\rho} \left[1 - \frac{1 - \dfrac{\mu}{\mu - s}A^*(s)}{1 - \dfrac{\mu}{\mu - s}} \right]$$

$$= \frac{1}{\rho} \left[1 - \frac{\mu - s - \mu A^*(s)}{-s} \right]$$

$$= \frac{1}{\mu\bar{t}} \left[\frac{-\mu[1 - A^*(s)]}{-s} \right]$$

or

$$\hat{I}^*(s) = \frac{1 - A^*(s)}{s\bar{t}}$$

which is the desired result.

(d) Although an idle period may never occur in the unstable queue, those that do occur will have durations properly described by the transforms discussed in Chapter 1. □

INDEX